普通高等学校
建筑环境与能源应用工程系列教材

# 建筑环境与能源应用工程专业英语（第2版）

Building Environment and Energy Engineering

主 编／白雪莲　刘　猛　李文杰
　　　　张慧玲　王建辉　牛润萍
主 审／龙惟定

重庆大学出版社

## 内容简介

本教材的课文和材料多来自英美学术团体设计手册,突出了英语的纯正性和知识的基础性,进一步提高学生掌握专业文献的英语表达和书写能力。全书共20课,内容涵盖了建筑环境与设备工程专业的所有专业课程,以及大数据、人工智能和工程伦理、职业道德。每课由5部分构成:第1部分为预备知识(preliminary),以图表的形式表现本课的主要内容;第2部分为课文(text),是教师重点讲解内容;第3部分为知识链接(extension),对课文中出现的比较重要的知识点进行扩充和解释;第4部分为词汇和表达(Words and Expressions),对课文中的专业词汇提供解释;第5部分为阅读材料(reading),围绕本课核心内容补充相关阅读材料,供学生课后拓展训练;第6部分为课后练习题(Exercise and Practice),包含练习题、翻译、写作及课后讨论等多种训练。

此外,教材还附录了本专业领域的部分知名国际学术团体和学术期刊、英文摘要写作和科技论文翻译,以培养学生查阅国外文献的能力,提高专业英语的基本应用能力。

**图书在版编目(CIP)数据**

建筑环境与能源应用工程专业英语/白雪莲,刘猛,李文杰,等主编. --2版. --重庆:重庆大学出版社,2019.8(2019.11重印)
普通高等学校建筑环境与能源应用工程系列教材
ISBN 978-7-5624-7687-0

Ⅰ.①建… Ⅱ.①白… ②刘… ③李… Ⅲ.①建筑工程—环境管理—英语—高等学校—教材 Ⅳ.①TU-023

中国版本图书馆 CIP 数据核字(2019)第182661号

普通高等学校建筑环境与能源应用工程系列教材
**建筑环境与能源应用工程专业英语**
(第2版)
白雪莲 刘 猛 李文杰 主编
张慧玲 王建辉 牛润萍
龙惟定 主审
责任编辑:林青山 版式设计:林青山
责任校对:万清菊 责任印制:张 策

\*

重庆大学出版社出版发行
出版人:饶帮华
社址:重庆市沙坪坝区大学城西路21号
邮编:401331
电话:(023)88617190 88617185(中小学)
传真:(023)88617186 88617166
网址:http://www.cqup.com.cn
邮箱:fxk@cqup.com.cn(营销中心)
全国新华书店经销
重庆升光电力印务有限公司印刷

\*

开本:787mm×1092mm 1/16 印张:18 字数:451千
2014年1月第1版 2019年8月第2版 2019年11月第3次印刷
印数:3 401—6 000
ISBN 978-7-5624-7687-0 定价:45.00元

本书如有印刷、装订等质量问题,本社负责调换
版权所有,请勿擅自翻印和用本书
制作各类出版物及配套用书,违者必究

# 编审委员会

顾　　问　田胜元(重庆大学)
　　　　　彦启森(清华大学)
　　　　　刘安田(中国人民解放军后勤工程学院)
主任委员　付祥钊(重庆大学)
委　　员　(排名按姓氏笔画)
　　　　　卢　军(重庆大学)
　　　　　付祥钊(重庆大学)
　　　　　安大伟(天津大学)
　　　　　李长惠(重庆大学出版社)
　　　　　李永安(山东建筑大学)
　　　　　刘光远(扬州大学)
　　　　　李　帆(华中科技大学)
　　　　　李安桂(西安建筑科技大学)
　　　　　连之伟(上海交通大学)
　　　　　张　旭(同济大学)
　　　　　张国强(湖南大学)
　　　　　吴祥生(中国人民解放军后勤工程学院)
　　　　　段常贵(哈尔滨工业大学)
　　　　　徐　明(中国建筑西南设计研究院)
　　　　　龚延风(南京工业大学)
　　　　　黄　晨(上海理工大学)
　　　　　裴清清(广州大学)
秘　　书　肖益民(重庆大学)
　　　　　张　婷(重庆大学出版社)

# 序

20世纪50年代初，为了北方采暖和工业厂房通风等迫切需要，全国在八所高校设立了"暖通"专业，随即增加了空调内容，培养以保障工业建筑生产环境、民用建筑生活与工作环境的本科专业人才。20世纪70年代末，又设立了燃气专业。1998年二者整合为"建筑环境与设备工程"。随后15年，全球能源环境形势日益严峻，而本专业在保障建筑环境上的能源消耗更是显著加大。保障建筑环境、高效应用能源成为当今社会对本专业的两大基本要求。2013年，国家再次扩展本专业范围，将建筑节能技术与工程、建筑智能设施二专业纳入，更名为"建筑环境与能源应用工程"。

本专业在内涵扩展的同时，规模也在加速发展。第一阶段，暖通燃气与空调工程阶段：近50年，本科招生院校由8所发展为68所；第二阶段，建筑环境与设备工程阶段：15年来，本科招生院校由68所发展到180多所，年招生规模达到1万人左右；第三阶段，建筑环境与能源应用工程阶段：这一阶段有多长，难以预见，但是本专业由配套工种向工程中坚发展是必然的。较之第二阶段，社会背景也有较大变化，建筑环境与能源应用工程必须面对全社会、全国和全世界的多样化人才需求。过去有利于学生就业和发展的行业与地方特色，现已露出约束毕业生人生发展的端倪。针对某个行业或地方培养人才的模式需要改变，本专业要实现的培养目标是建筑环境与能源应用工程专业的复合型工程技术应用人才。这样的人才是服务于全社会的。

本专业科学技术的新内容主要在能源应用上：重点不是传统化石能源的应用，而是太阳辐射和存在于空气、水体、岩土等环境中的可再生能源的应用；应用的基本方式不再局限于化石燃料燃烧产生的热能，将是依靠动力从环境中采集与调整热能；应用的核心设备不再是锅炉，将是热泵。专业工程实践方面：传统领域即设计与施工仍需进一步提高；新增的工作将是从城市、城区、园区到建筑四个层次的能源需求的预测与保障、规划与实施，从工程项目的策划立项、方案制订、设计施工到运行使用全过程提高能源应用效率，从单纯的能源应用技术到综合的能源管理等。这些急需开拓的成片的新领域，也是本专业与热能动力专业在能源应用上的主要区别。本专业将在能源环境的强约束下，满足全社会对人居建筑环境和生产工艺环境提出的新需求。

本专业将不断扩展视野，改进教育理念，更新教学内容和教学方法，提升专业教学水平。将在建筑环境与设备工程专业的基础上，创建特色课程，完善专业知识体系。专业基础部分包括建筑环境学、流体力学、工程热力学、传热学、热质交换原理与设备、流体输配管网等理论知识；专业部分包括室内环境控制系统、燃气储存与输配、冷热源工程、城市燃气工程、城市能源规划、建筑能源管理、工程施工与管理、建筑设备自动化、建筑环境测试技术等系统的工程技术知识。各校需要结合自己的条件，设置相应的课程体系，使学生建立起有自己特色的专业知识体系。

本专业知识体系由知识领域、知识单元以及知识点三个层次组成，每个知识领域包含若干个知识单元，每个知识单元包含若干知识点，知识点是本专业知识体系的最小集合。课程设置不能割裂知识单元，并要在知识领域上加强关联，进而形成专业的课程体系。重庆大学出版社积极学习了解本专业的知识体系，针对重庆大学和其他高校设置的本专业课程体系，规划出版建筑环境与能源应用工程专业系列教材，组织专业水平高、教学经验丰富的教师编写。

这套专业系列教材口径宽阔、核心内容紧凑，与课程体系密切衔接，便于教学计划安排，有助于提高学时利用效率。学生通过这套系列教材的学习，能够掌握建筑环境与能源应用领域的专业理论、设计和施工方法。结合实践教学，还能帮助学生熟悉本专业施工安装、调试与试验的基本方法，形成基本技能；熟悉工程经济、项目管理的基本原理与方法；了解与本专业有关的法规、规范和标准，了解本专业领域的现状和发展趋势。

这套系列教材，还可用于暖通、燃气工程技术人员的继续教育。对那些希望进入建筑环境与能源应用工程领域发展的其他专业毕业生，也是很好的自学课本。

这是对建筑环境与能源应用工程系列教材的期待！

2013 年 5 月于重庆大学虎溪校区

# 前 言
(第 2 版)

本教材第 1 版经过了 5 年的教学实践,得到了广大师生的应用反馈。同时,建筑环境与能源应用工程专业领域随着科技进步和社会发展也在不断拓展。为此,编写组在第 1 版的基础上对教材内容进行了以下考虑和改进:

1. 大数据、智能化等概念已深入各行各业。增加 Lesson 19(Big Data in HVAC),以保证本教材与前沿领域的紧密联系。增加的 Lesson 20(Engineering Ethics and Professional Morality)体现了高校教育人才培养中"立德树人"的重要理念。

2. 鉴于本科毕业设计(论文)学习阶段,有专业英文文献翻译和英文摘要写作的要求,增加英文摘要写作和科技论文翻译两部分附录,从而提高学生专业英语的基本应用能力。

3. 专业英文文献的阅读对专业素养和英文技能的提高都大有裨益。因此,为了引导学生阅读课外文献,每课都增加了对应主题内容的数字资源。

4. 为了方便教与学,Part 1 Preliminary 增加对图示的英文解释,Part 2 Text 增加对课文正文的中文概述,增加 Part 4 Words and Expressions。课后练习题(Part 6 Exercise and Practice)在原有 Discussion 的练习题之外,增加 Translation/Writing 等多种练习题。

本版教材第 1,2,13 课由王建辉(重庆交通大学)负责编写,第 3,4,12 课由李文杰(重庆科技学院)负责编写,第 5,6,16,17,20 课由刘猛(重庆大学)和牛润萍(北京建筑大学)负责编写,第 7,8,9,15,19 课由白雪莲(重庆大学)负责编写,第 10,11,14,18 课由张慧玲(重庆交通大学)负责编写。

限于编者经验和水平,教材中难免存在不妥之处,敬请各位读者提出宝贵意见。

编 者
2019 年 5 月

# 前 言
(第1版)

随着经济全球化、教育国际化的发展,对既精通专业知识又精通外语的高素质复合型人才的需求日益旺盛。加强专业英语的学习是高等学校人才培养主动适应社会需求,积极推动高等教育国际化进程的主要措施之一。建筑环境与能源应用工程专业名称的调整,使得专业教学内容和培养目标由原来的建筑暖通向环境调控和能源应用领域拓展,从而对专业英语的教材内容也提出了新的要求。

本教材的特点主要体现在以下4个方面:

(1)内容。基本上涵盖了建筑环境与能源应用工程专业的所有专业基础课程和专业课程内容。避免语言学习与专业知识的脱节,使学生更好地掌握专业课程知识,同时提高专业英语应用能力。

(2)难度。本教材主要面对建筑环境与能源应用工程专业及相关专业的大学本科和专科学生,英语生词量和语法结构上基本符合这一级别的要求,教材内容以基础知识为主,与中文专业课程的学习相辅相成。

(3)选材。课文和阅读材料多来自广受专业接受和应用的英美学术团体设计手册,更加突出了英语的纯正性和知识的基础性,进一步提高学生专业科技英语的表达和书写能力。

(4)编排。与以往的专业英语教材有所不同,每课由5部分构成:第1部分为预备知识(preliminary),以图表的形式表现本课的主要内容。第2部分为课文(text),是教师重点讲解内容,词汇与术语采用直接备注的形式,保证阅读的连贯性;并在正文后列出关键词(keywords),以便学生据此搜索相关文献,扩展阅读。第3部分为知识链接(extension),即将课文中出现的对本专业比较重要的知识点,进行扩充和解释。第4部分为阅读材料(reading),围绕本课核心内容补充相关阅读材料,供学生课后拓展训练。第5部分为课后讨论(discussion),就本课的主要内容引导学生思考和讨论。

此外,教材还附录了本专业领域的部分知名国际学术团体和学术期刊,以培养学生阅读外文文献或访问相关外文网站的兴趣和能力,为学生进一步开展相关领域的学习和科研打下基础。

全书编写分工如下:第1,2,13课由王建辉(重庆交通大学)负责编写,第3,4,12课由李文杰(重庆科技学院)负责编写,第5,6,16,17课由刘猛(重庆大学)和牛润萍(北京建筑工程大学)负责编写,第7,8,9,15课由白雪莲(重庆大学)负责编写,第10,11,14,18课由张慧玲(重庆交通大学)负责编写。

同济大学龙惟定教授对书稿进行了认真审校,并提出了许多宝贵意见。此外,本次编写得到了重庆大学出版社的大力支持,特别是在几年前出版社有此想法之初,就不懈努力,催促行动。正是有了他们的坚持,才使得本书最终得以付梓。

本书可作为建筑环境与能源应用工程专业及其相关专业本、专科学生的专业英语教材,任课教师可根据具体要求选择所需内容。同时还可供上述专业的教师、科研人员和工程技术人员参考。

由于编者水平有限,书中难免有不妥之处,衷心欢迎广大师生指正。

编　者

2013年8月

# Contents

**Lesson 1  Thermodynamics and Heat Transfer** ................................................ 1

    Part 1   Preliminary ................................................................................ 1
    Part 2   Text ............................................................................................ 2
    Part 3   Extension .................................................................................. 8
    Part 4   Words and Expressions ........................................................ 9
    Part 5   Reading .................................................................................... 10
    Part 6   Exercises and Practices ........................................................ 13

**Lesson 2  Hydrodynamics and Fluid Machinery** ............................................ 15

    Part 1   Preliminary ................................................................................ 15
    Part 2   Text ............................................................................................ 16
    Part 3   Extension .................................................................................. 23
    Part 4   Words and Expressions ........................................................ 24
    Part 5   Reading .................................................................................... 26
    Part 6   Exercises and Practices ........................................................ 28

**Lesson 3  Thermal Comfort** ................................................................................ 29

    Part 1   Preliminary ................................................................................ 29
    Part 2   Text ............................................................................................ 29
    Part 3   Extension .................................................................................. 35
    Part 4   Words and Expressions ........................................................ 35
    Part 5   Reading .................................................................................... 36
    Part 6   Exercise and Practice ............................................................ 41

| | | | |
|---|---|---|---|
| **Lesson 4** | **Indoor Environmental Health** | ············· | 43 |
| | Part 1　Preliminary ············· | | 43 |
| | Part 2　Text ············· | | 43 |
| | Part 3　Extension ············· | | 49 |
| | Part 4　Words and Expressions ············· | | 49 |
| | Part 5　Reading ············· | | 50 |
| | Part 6　Exercise and Practice ············· | | 56 |
| **Lesson 5** | **Introduction of Heating System** ············· | | 57 |
| | Part 1　Preliminary ············· | | 57 |
| | Part 2　Text ············· | | 58 |
| | Part 3　Extension ············· | | 63 |
| | Part 4　Words and Expressions ············· | | 63 |
| | Part 5　Reading ············· | | 64 |
| | Part 6　Exercises and Practices ············· | | 67 |
| **Lesson 6** | **Heating System Design** ············· | | 69 |
| | Part 1　Preliminary ············· | | 69 |
| | Part 2　Text ············· | | 69 |
| | Part 3　Extension ············· | | 77 |
| | Part 4　Words and Expressions ············· | | 78 |
| | Part 5　Reading ············· | | 79 |
| | Part 6　Exercises and Practices ············· | | 82 |
| **Lesson 7** | **Air Conditioning Fundamentals** ············· | | 84 |
| | Part 1　Preliminary ············· | | 84 |
| | Part 2　Text ············· | | 84 |
| | Part 3　Extension ············· | | 89 |
| | Part 4　Words and Expressions ············· | | 90 |
| | Part 5　Reading ············· | | 91 |
| | Part 6　Exercises and Practices ············· | | 97 |
| **Lesson 8** | **Air Conditioning Systems** ············· | | 98 |
| | Part 1　Preliminary ············· | | 98 |
| | Part 2　Text ············· | | 98 |
| | Part 3　Extension ············· | | 104 |
| | Part 4　Words and Expressions ············· | | 104 |
| | Part 5　Reading ············· | | 105 |

|  |  |  |  |
|---|---|---|---|
| | Part 6 | Exercises and Practices | 110 |
| **Lesson 9** | **Control Systems for HVAC** | | 112 |
| | Part 1 | Preliminary | 112 |
| | Part 2 | Text | 112 |
| | Part 3 | Extension | 118 |
| | Part 4 | Words and Expressions | 118 |
| | Part 5 | Reading | 119 |
| | Part 6 | Exercises and Practices | 125 |
| **Lesson 10** | **Ventilation System** | | 126 |
| | Part 1 | Preliminary | 126 |
| | Part 2 | Text | 127 |
| | Part 3 | Extension | 130 |
| | Part 4 | Words and Expressions | 131 |
| | Part 5 | Reading | 132 |
| | Part 6 | Exercises and Practices | 136 |
| **Lesson 11** | **Ventilation Application** | | 137 |
| | Part 1 | Preliminary | 137 |
| | Part 2 | Text | 138 |
| | Part 3 | Extension | 144 |
| | Part 4 | Words and Expressions | 144 |
| | Part 5 | Reading | 145 |
| | Part 6 | Exercises and Practices | 149 |
| **Lesson 12** | **An Introduction to Cleanroom** | | 151 |
| | Part 1 | Preliminary | 151 |
| | Part 2 | Text | 151 |
| | Part 3 | Extension | 156 |
| | Part 4 | Words and Expressions | 156 |
| | Part 5 | Reading | 157 |
| | Part 6 | Exercise and Practice | 162 |
| **Lesson 13** | **Refrigeration System** | | 164 |
| | Part 1 | Preliminary | 164 |
| | Part 2 | Text | 165 |
| | Part 3 | Extension | 171 |
| | Part 4 | Words and Expressions | 172 |

|  |  |  |  |
|---|---|---|---|
| | Part 5 | Reading | 172 |
| | Part 6 | Exercises and Practices | 177 |

**Lesson 14  Heat Pump** ......... 178
    Part 1    Preliminary ......... 178
    Part 2    Text ......... 179
    Part 3    Extension ......... 186
    Part 4    Words and Expressions ......... 186
    Part 5    Reading ......... 188
    Part 6    Exercises and Practices ......... 191

**Lesson 15  Energy Resource** ......... 192
    Part 1    Preliminary ......... 192
    Part 2    Text ......... 192
    Part 3    Extension ......... 197
    Part 4    Words and Expressions ......... 198
    Part 5    Reading ......... 199
    Part 6    Exercises and Practices ......... 204

**Lesson 16  Introduction of Fuel Gas** ......... 205
    Part 1    Preliminary ......... 205
    Part 2    Text ......... 206
    Part 3    Extension ......... 212
    Part 4    Words and Expressions ......... 212
    Part 5    Reading ......... 213
    Part 6    Exercises and Practices ......... 216

**Lesson 17  Gas Transportation Process** ......... 217
    Part 1    Preliminary ......... 217
    Part 2    Text ......... 217
    Part 3    Extension ......... 223
    Part 4    Words and Expressions ......... 223
    Part 5    Reading ......... 224
    Part 6    Exercises and Practices ......... 226

**Lesson 18  Building Energy Efficiency** ......... 227
    Part 1    Preliminary ......... 227
    Part 2    Text ......... 228
    Part 3    Extension ......... 235

|  |  |  |
|---|---|---|
| Part 4 | Words and Expressions | 235 |
| Part 5 | Reading | 237 |
| Part 6 | Exercises and Practices | 240 |

**Lesson 19　Big Data in HVAC** ······ 242

|  |  |  |
|---|---|---|
| Part 1 | Preliminary | 242 |
| Part 2 | Text | 242 |
| Part 3 | Extension | 247 |
| Part 4 | Words and Expressions | 247 |
| Part 5 | Reading | 249 |
| Part 6 | Exercise and Practice | 253 |

**Lesson 20　Engineering Ethics and Professional Morality** ······ 254

|  |  |  |
|---|---|---|
| Part 1 | Preliminary | 254 |
| Part 2 | Text | 254 |
| Part 3 | Extension | 257 |
| Part 4 | Words and Expressions | 257 |
| Part 5 | Reading | 258 |
| Part 6 | Exercise and Practice | 259 |

**附　录** ······ 261

附录1　国际专业学术团体 ······ 261
附录2　国际专业学术期刊 ······ 263
附录3　英文摘要写作 ······ 264
附录4　科技论文翻译（中文） ······ 267

**References and Sources** ······ 271

# Lesson 1

# Thermodynamics and Heat Transfer

## Part 1  Preliminary

A thermodynamic cycle is a series of thermodynamic processes which returns a system to its initial state. Properties depend only on the thermodynamic state and thus do not change over a cycle. Variables such as heat and work are not zero over a cycle, but rather depend on the process. A heat engine (Fig. 1.1) is a continuous cyclic device that produces positive network output by adding heat.

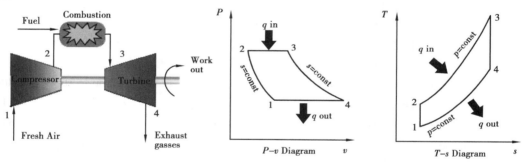

Fig. 1.1  **A thermodynamic cycle**

Heat is transferred between any two bodies by one or more of three modes: conduction, convection, and radiation. Thermal conduction (Fig. 1.2) refers to the direct transfer of energy between particles at the atomic level. Thermal convection (Fig. 1.3) may include some conduction but refers primarily to energy transfer by eddy mixing and diffusion, i.e., by fluids in motion. Thermal radiation (Fig. 1.4) describes a complex phenomenon which includes changes in energy form: from internal energy at the source to electromagnetic energy for transmission, then back to internal energy at the receiver.

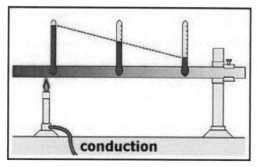

Fig. 1.2  **Conduction by a metal bar**

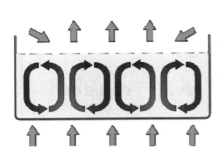

Fig. 1.3　Convection cells in a gravity field

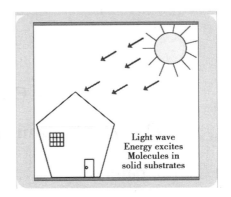

Fig. 1.4　Radiation by solar

# Part 2　Text

热力学是一门研究物质的能量、能量传递和转换以及能量与物质性质之间普遍关系的学科。工程热力学是热力学的工程分支,是在阐述热力学普遍原理的基础上,研究这些原理的技术应用学科。本章节着重介绍热量与其他形式能量(主要是机械能)之间的转换规律及应用,详细介绍了热力学的概念,并对热力学第一定律、热力学第二定律以及相关的热力系统进行了较为详细的阐述和讨论。热力学第三定律将在本章第三部分以中文形式给出。

Thermodynamics is the study of energy, its transformations, and its relation to states of matter. This text covers the basic definitions and the application of thermodynamics. The first part reviews the first and second law of thermodynamics and present methods for calculating thermodynamic properties. The second part simply addresses the thermodynamics system. The Third Law of Thermodynamics[①] will be given in Chinese.

## 1. *Thermodynamics*

Thermodynamics is the branch of physical science that deals with the various phenomena of energy and related properties of matter, especially of the laws of transformations of heat into other forms of energy and vice-versa. A thermodynamic system is a region in space or a quantity of matter bounded by a closed surface. The surroundings include everything external to the system, and the system is separated from the surroundings by the system boundaries. These boundaries can be movable or fixed, real or imaginary.

The concepts that operate in any thermodynamic system are entropy and energy. Entropy measures the molecular disorder of a system. The more mixed a system, the greater its entropy; conversely, an orderly or unmixed configuration is one of low entropy. Energy has the capacity for producing an effect and can be categorized either stored or transient forms as described in the following sections:

---

① 见 Part 3 Extension.

## Stored forms of energy include:

Thermal (internal) energy, $u$—the energy possessed by a system caused by the motion of the molecules and/or intermolecular forces.

potential energy, $P.E.$—the energy possessed by a system caused by the attractive forces existing between molecules, or the elevation of the system:

$$P.E. = mgz \tag{1.1}$$

Where,

$m$ = mass;

$g$ = local acceleration of gravity;

$z$ = elevation above a horizontal reference plane.

kinetic energy, $K.E.$—the energy possessed by a system caused by the velocity of the molecules:

$$K.E. = \frac{mv^2}{2} \tag{1.2}$$

Where,

$m$ = mass;

$v$ = velocity of the fluid streams crossing system boundaries.

Chemical energy, $Ec$—energy possessed by the system caused by the arrangement of atoms composing the molecules.

Nuclear (atomic) energy, $Ea$—energy possessed by the system from the cohesive forces holding protons and neutrons together as the atom's nucleus.

## Transient energy forms include:

Heat ($Q$) is the mechanism that transfers energy across the boundary of systems with differing temperatures, always in the direction of the lower temperature. Heat is positive when energy is added to the system.

Work is the mechanism that transfers energy across the boundary of systems with differing pressures (or force of any kind), always in the direction of the lower pressure; if the total effect produced in the system can be reduced to the raising of a weight, then nothing but work has crossed the boundary.

Mechanical or shaft work ($W$) is the energy delivered or absorbed by a mechanism, such as a turbine, air compressor or internal combustion engine.

Flow work is energy carried into or transmitted across the system boundary because a pumping process occurs somewhere outside the system, causing fluid to enter the system. It can be more easily understood as the work done by the fluid just outside the system on the adjacent fluid entering the system to force or push it into the system. Flow work also occurs as fluid leaves the system.

$$\text{Flow Work (per unit mass)} = pv \tag{1.3}$$

Where $p$ is the pressure and $v$ is the specific volume, or the volume displaced per unit mass.

A property of a system is any observable characteristic of the system. The state of a system is defined by listing its properties. The most common thermodynamic properties are: temperature $t$, pressure $p$, and specific volume $v$ or density $\rho$. Additional thermodynamic properties include entropy, stored forms of energy and enthalpy.

Frequently, thermodynamic properties combine to form new properties. Enthalpy ($h$), a result of combining properties, is defined as:

$$h = u + pv \tag{1.4}$$

Where,

$u$ = internal energy per unit mass;

$p$ = pressure;

$v$ = specific volume.

Each property in a given state has only one definite value, and any property always has the same value for a given state, regardless of how the substance arrived at that state.

A process is a change in state that can be defined as any change in the properties of a system. A process is described by specifying the initial and final equilibrium states, the path (if identifiable) and the interactions that take place across system boundaries during the process.

A cycle is a process, or more frequently, a series of processes wherein the initial and final states of the system are identical. Therefore, at the conclusion of a cycle all the properties have the same value they had at the beginning.

A pure substance has a homogeneous and invariable chemical composition. It can exist in more than one phase, but the chemical composition is the same in all phases.

If a substance exists as vapor at the saturation temperature, it is called saturated vapor. (Sometimes the term dry saturated vapor is used to emphasize that the quality is 100%.) When the vapor is at a temperature greater than the saturation temperature, it is superheated vapor. The pressure and temperature of superheated vapor are independent properties, since the temperature can increase while the pressure remains constant. Gases are highly superheated vapors.

## The First Law of Thermodynamics

The First Law of Thermodynamics plays a significant role in the analysis of thermodynamic systems, and for this reason we shall devote considerable attention to its development. However, it is useful for us to outline briefly the essential features of the first law of thermodynamics as it applies to closed systems. With this information, we will be better prepared to organize and interpret the details of the succeeding development.

In essence, the First Law of Thermodynamics is a generalization of the observed facts about the energy interactions between a system and its environment. Specifically, it relates the various energy interactions between a system and its environment to changes of state experienced by that system during these interactions. All physical experience has confirmed that the generalizations embodied in the first law of thermodynamics must be satisfied to describe realistically a physical situation.

There are many ways in which the first law of thermodynamics can be stated, and some rather complex arguments are required to prove the equivalence of these statements. However, the simplest of these statements can be formulated by considering a system which executes a cycle. By definition, the system experiences no net change of state. Therefore, there can be no net energy interaction between the system and its environment. For our purposes it will be sufficient to consider only two forms of energy interaction between a system and its environment: work transfers $W$, and heat transfers $Q$. Thus, we may state the first law of thermodynamics in a formal way as:

The net energy interaction between a system and its environment is zero for a cycle executed by the system.

The mathematical equivalent of this statement is:

$$\oint \delta Q - \oint \delta W = 0 \tag{1.5}$$

The integral sign indicates the algebraic summation of each infinitesimal heat transfer $\delta Q$ or work transfer $\delta W$ over the complete cycle. The negative sign is introduced because historically in thermodynamics, work transfer from a system is taken as positive which is the inverse of the sign convention used in mechanics.

The major difficulty in applying the first law to physical situations lies in the determination and formulation of the heat transfers and work transfers. This will soon become apparent as we take the basic concepts of work transfer directly from mechanics and electro mechanics, and reformulate them from a thermodynamics point of view. However, as we deal with progressively more complex systems, it may be necessary to extend these definitions of work transfer to avoid any ambiguity. Since heat transfer is unique to thermodynamics, it is not possible to rely on other disciplines for its formulation. In fact, one of the most difficult aspects of the formalism of thermodynamics is a rigorous definition of heat transfer. Therefore, we shall delay its discussion until later in our development. Presently, it is sufficient to regard heat transfer simply as one method of energy interaction, distinct from work transfer, between a system and its environment.

As we shall see later, it will become increasingly important for us to distinguish between the energy transfer interactions $Q$ and $W$ and the property which changes by virtue of these interactions. This latter quantity is defined as the energy, $E$, of the system at a given state. It is simply the stored energy of the system. The change in the stored energy of a system due to a change of state of the system from state 1 to state 2 is defined by

$$\int_1^2 \delta Q - \int_1^2 \delta W = E_2 - E_1 \tag{1.6}$$

## The Second Law of Thermodynamics

The Second Law of Thermodynamics is capable of indicating the maximum possible efficiencies of heat engines, coefficient of performance of heat pumps and refrigerators, defining a temperature scale independent of physical properties etc.

Heat reservoir is the system having very large heat capacity i.e. it is a body capable of absorbing or rejecting finite amount of energy without any appreciable change in its' temperature. Thus in general it may be considered as a system in which any amount of energy may be dumped or extracted out and there shall be no change in its temperature. Large river, sea etc. can also be considered as reservoir, as dumping of heat to it shall not cause appreciable change in temperature.

Heat reservoirs can be of two types depending upon nature of heat interaction i. e. heat rejection or heat absorption from it. Heat reservoir which rejects heat from it is called source. While the heat reservoir which absorbs heat is called sink. Sometimes these heat reservoirs may also be called Thermal Energy Reservoirs (TER).

Heat engine is a device used for converting heat into work as it has been seen from nature that conversion from work to heat may take place easily but the vice-versa is not simple to be realized. Heat and work have been categorized as two forms of energy of low grade and high grade type. Conversion of high grade of energy to low grade of energy may be complete (100%), and can occur directly whereas complete conversion of low grade of energy into high grade of energy is not possible. For converting low grade of energy (heat) into high grade of energy (work) some device called heat engine is required.

Thus, heat engine may be precisely defined as "a device operating in cycle between high temperature source and low temperature sink and producing work". Heat engine receives heat from source, transforms some portion of heat into work and rejects balance heat to sink. All the processes occurring in heat engine constitute cycle.

Block diagram representation of a heat engine is shown as Fig. 1.5. A practical arrangement used in gas turbine plant is also shown for understanding the physical significance of heat engine.

Gas turbine installation (Fig. 1.6) shows that heat is added to working fluid from 1-2 in a "heat exchanger 1" and may be treated as heat supply by source. Working fluid is expanded in turbine from 2-3 and produces positive work. After expansion fluid goes to the "heat exchanger 2" where it rejects heat from it like heat rejection in sink. Fluid at state 4 is sent to compressor for being compressed to state 1. Work required for compression is quite small as compared to positive work available in turbine and is supplied by turbine itself. As the show of Fig. 1.7.

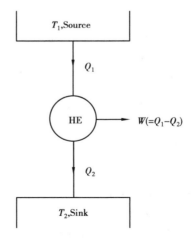

Fig. 1.5  **Heat engine**

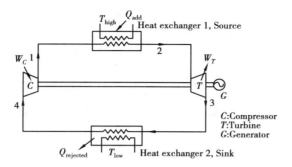

Fig. 1.6  Closed cycle gas turbine power plant

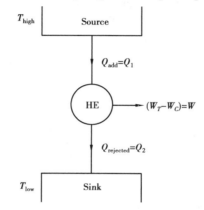

Fig. 1.7  Heat engine representation for gas turbine plant

## 2. *Thermodynamic system*

In the engineering world, objects normally are not isolated from one another. In most engineering problems many objects enter into a given problem. Some of these objects, all of these objects, or even additional ones may enter into a second problem. The nature of a problem and its solution are dependent on which objects are under consideration. Thus, it is necessary to specify which objects are under consideration in a particular situation. **In thermodynamics this is done either by placing an imaginary envelope around the objects under consideration or by using an actual envelope if such exists.** [2] The term system refers to everything lying inside the envelope. The envelope, real or imaginary, is referred to as the boundaries of the system. It is essential that the boundaries of the system be specified very carefully. For example, when one is dealing with a gas in a cylinder where the boundaries are located on the outside of the cylinder, the system includes both the cylinder and its contained gas. On the other hand, when the boundaries are placed at the inner face of the cylinder, the system consists solely of the gas itself.

When the boundaries of a system are such that it cannot exchange matter with the surroundings, the system is said to be a closed system. The system, however, may exchange energy in the form of heat or work with the surroundings. The boundaries of a closed system may be rigid or may expand

---

② 见 Part 3 Extension.

or contract, but the mass of a closed system cannot change. Hence, the term control mass sometimes is used for this type of system. When the energy crossing the boundaries of a closed system is zero or substantially so, the system may be treated as an isolated system.

In most engineering problems, matter, generally a kind of fluid crosses the boundaries of a system in one or more places. Such a system is known as an open system. The boundaries of an open system are so placed that their location does not change with time. Thus, the boundaries enclose a fixed volume, commonly known as the control volume.

Frequently the total system to be considered may be large and complicated. The system may be broken down into component parts and an analysis of the component parts made. Then the performance of the entire system can be determined by the summation of the performance of the individual component system. For example, consider the liquid-vapor part of a steam power plant as an entire system. This system, which is closed, contains the steam generator, the steam turbine, the steam condenser, the feed-water pumps, and the feed-water heaters. Since a fluid enters and leaves each of these smaller systems, each one is an open system and must be analyzed as such.

**Keywords**: thermal energy, kinetic energy, flow work, thermodynamic system

**Source from**:

[1] 张寅平,潘毅群,王馨. 建筑环境与设备工程专业. 第一课 THERMODYNAMICS AND REFRIGERATION CYCLES. 北京:中国建筑工业出版社, 2005.

[2] 赵三元,闫岫峰. 建筑类专业英语——暖通与燃气. 第一课 BASIC CONCEPTS AND DEFINITIONS. 第二课 APPLICATION OF THE PRINCIPLES OF THERMODYNAMICS. 北京:中国建筑工业出版社, 2002.

[3] Onkar Singh. Applied Thermodynamics. 3rd Edition. Chapter 3: First Law of Thermodynamics; Chapter4: Second Law of Thermodynamics. New Age International Ltd, publishers, 2009.

# Part 3　Extension

1. The Third Law of Thermodynamics(热力学第三定律)

热力学第三定律是对熵的论述:一般当封闭系统达到稳定平衡时,熵应该为最大值,在任何过程中,熵总是增加,但理想气体如果是绝热可逆过程,熵的变化为零。但是理想气体实际并不存在,所以现实物质中,即使是绝热可逆过程,系统的熵也在增加,不过增加得少。在绝对零度,任何完美晶体的熵为零。

2. In thermodynamics this is done either by placing an imaginary envelope around the objects under consideration or by using an actual envelope of such exists.

在热力学中,系统的边界是可以实际存在的,也可以是假想的,边界的选取取决于研究的对象和热力过程。另外,系统和外界的边界可以是固定不动的,也可以是有位移和形变的。

3. With the valves closed, the system is a closed one. However, with either or both of the valves open, the system becomes an open system.

一个热力系统如果和外界只有能量交换而无物质交换,则该系统称为闭口系统,又称闭口系。如果热力系统和外界不仅有能量交换而且有物质交换,则该系统称为开口系统,又称开口系。在本文此处,如果阀门关闭,系统就是闭口系统。但是,如果两个阀门都关闭或者关闭其中的一个,该系统就是开口系统。区分开口系统和闭口系统的关键是边界上是否有物质通过。

## Part 4  Words and Expressions

| | |
|---|---|
| thermodynamics | n. 热力学 |
| property | n. 性质,性能 |
| phenomena | n. 现象 |
| entropy | n. [热]熵(热力学函数) |
| enthalpy | n. [热]焓 |
| configuration | n. 配置;结构;外形 |
| categorize | vt. 分类 |
| mass | n. 质量 |
| thermal energy | 热能 |
| potential energy | 势能 |
| kinetic energy | 动能 |
| local acceleration of gravity | 当地重力加速度 |
| boundary | n. 边界;范围;分界线 |
| adjacent | adj. 邻近的,毗连的 |
| characteristic | n. 特征;特性;特色 |
| specific volume | 比容;比体积 |
| equilibrium | n. 均衡;平静;保持平衡的能力 |
| substance | n. 物质;实质 |
| homogeneous | adj. 均匀的;同质的 |
| composition | n. 构成;合成物;成分 |
| saturated vapor | 饱和蒸汽 |
| superheated | adj. 过热的,过热蒸汽 |
| equivalence | n. 等值;相等 |
| integral | adj. 积分的;整体的 |
| algebraic | adj. [数]代数的;关于代数学的 |
| infinitesimal | adj. 无穷小的;无限小的;极小的 |
| positive | adj. [数]正的 |
| negative | adj. [数]负的 |
| sign convention | 符号法则 |
| electro | n. 电镀物品 |
| ambiguity | n. 含糊;不明确 |
| formalism | n. 形式体系 |
| coefficient of performance | 性能系数 |
| heat pump | 热泵 |
| refrigerator | n. 冰箱,冷藏库 |
| extract | vt. 提取;取出 |

续表

| reservoir | *n.* 水库；蓄水池 |
| appreciable | *adj.* 可评估的；相当可观的 |
| conversion | *n.* 转换；变换 |
| compressor | *n.* 压缩机 |
| grade | *n.* 等级；级别 |
| cylinder | *n.* 圆筒；汽缸 |

# Part 5　Reading

## Heat Transfer Modes

Heat is transferred between any two bodies by one or more of three modes: conduction, convection, and radiation. Thermal conduction refers to the direct transfer of energy between particles at the atomic level. Thermal convection may include some conduction but refers primarily to energy transfer by eddy mixing and diffusion, i.e., by fluids in motion. Thermal radiation describes a complex phenomenon which includes changes in energy form: from internal energy at the source to electromagnetic energy for transmission, then back to internal energy at the receiver. Radiation transfer requires no intervening material, and in fact works best in a perfect vacuum. In accordance with the second law of thermodynamics, net heat transfer occurs in the direction of decreasing temperature.

## 1. *Conduction*

For steady-state conduction in one direction through a homogeneous material, the Fourier equation applies:

$$q = -kA\frac{\mathrm{d}t}{\mathrm{d}x} \tag{1.7}$$

Where,

　　$q$—heat transfer rate;

　　$k$—thermal conductivity;

　　$A$—area normal to flow;

　　$\dfrac{\mathrm{d}t}{\mathrm{d}x}$—temperature gradient.

The minus sign shows that heat flow takes place from a higher to a lower temperature.

### Steady-state conduction

Steady-state conduction is the form of conduction that happens when the temperature difference(s) driving the conduction are constant, so that (after an equilibration time), the spatial distribution of temperatures (temperature field) in the conducting object does not change any further. Thus, all partial derivatives of temperature with respect to space may either be zero or have nonzero values, but all derivatives of temperature at any point with respect to time are uniformly zero. In steady state conduction, the amount of heat entering any region of an object is equal to amount of heat coming out (if this were not so, the temperature would be rising or falling, as thermal energy was tapped or trapped in a region).

For example, a bar may be cold at one end and hot at the other, but after a state of steady state conduction is reached, the spatial gradient of temperatures along the bar does not change any further, as time proceeds. Instead, the temperature at any given section of the rod remains constant, and this temperature varies linearly in space, along the direction of heat transfer.

In steady state conduction, all the laws of direct current electrical conduction can be applied to "heat currents". In such cases, it is possible to take "thermal resistances" as the analog to electrical resistances. In such cases, temperature plays the role of voltage, and heat transferred per unit time (heat power) is the analog of electrical current. Steady state systems can be modeled by networks of such thermal resistances in series and in parallel, in exact analogy to electrical networks of resistors. See purely resistive thermal circuits for an example of such a network.

## Transient conduction

In general, during any period in which temperatures are changing in time at any place within an object, the mode of thermal energy flow is termed transient conduction. Another term is "non steady-state" conduction, referring to time-dependence of temperature fields in an object. Non-steady-state situations appear after an imposed change in temperature at a boundary of an object. They may also occur with temperature changes inside an object, as a result of a new source or sink of heat suddenly introduced within an object, causing temperatures near the source or sink to change in time.

When a new perturbation of temperature of this type happens, temperatures within the system will change in time toward a new equilibrium with the new conditions, provided that these do not change. After equilibrium, heat flow into the system will once again equal the heat flow out, and temperatures at each point inside the system no longer change. Once this happens, transient conduction is ended, although steady-state conduction may continue if there continues to be heat flow.

If changes in external temperatures or internal heat generation changes are too rapid for equilibrium of temperatures in space to take place, then the system never reaches a state of unchanging temperature distribution in time, and the system remains in a transient state.

An example of a new source of heat "turning on" within an object which causes transient conduction, is an engine starting in an automobile. In this case the transient thermal conduction phase for the entire machine would be over, and the steady state phase would appear, as soon as the engine had reached steady-state operating temperature. In this state of steady-state equilibrium, temperatures would vary greatly from the engine cylinders to other parts of the automobile, but at no point in space within the automobile would temperature be increasing or decreasing. After establishment of this state, the transient conduction phase of heat transfer would be over.

New external conditions also cause this process: for example the copper bar in the example steady-state conduction would experience transient conduction as soon as one end was subjected to a different temperature from the other. Over time, the field of temperatures inside the bar would reach a new steady-state, in which a constant temperature gradient along the bar will finally be set up, and this gradient would then stay constant in space. Typically, such a new steady state gradient is approached exponentially with time after a new temperature-or-heat source or sinks, has been introduced. When a "transient conduction" phase is over, heat flow may still continue at high power, so long as temperatures do not change.

## 2. *Convection*

Convective heat transfer is a mechanism of heat transfer occurring because of bulk motion of fluids. Heat is the entity of interest being adverted, and diffused. This can be contrasted with conductive heat transfer, which is the transfer of energy by vibrations at a molecular level through a solid or fluid, and irradiative heat transfer, the transfer of energy through electromagnetic waves.

Heat is transferred by convection in numerous examples of naturally occurring fluid flow, such as: wind, oceanic currents, and movements within the Earth's mantle. Convection is also used in engineering practices to provide desired temperature changes, as in heating of homes, industrial processes, cooling of equipment, etc.

### Forced convection

In forced convection, also called heat advection, fluid movement results from external surface forces such as a fan or pump. Forced convection is typically used to increase the rate of heat exchange. Many types of mixing also utilize forced convection to distribute one substance within another. Forced convection also occurs as a by-product to other processes, such as the action of a propeller in a fluid or aerodynamic heating. Fluid radiator systems, and also heating and cooling of parts of the body by blood circulation, are other familiar examples of forced convection.

Forced convection may happen by natural means, such as when the heat of a fire causes expansion of air and bulk air flow by this means. In microgravity, such flow (which happens in all directions) along with diffusion is the only means by which fires are able to draw in fresh oxygen to maintain themselves. The shock wave that transfers heat and mass out of explosions is also a type of forced convection.

Although forced convection from thermal gas expansion in zero-g does not fuel a fire as well as natural convection in a gravity field, some types of artificial forced convection are far more efficient than free convection, as they are not limited by natural mechanisms. For instance, a convection oven works by forced convection, as a fan which rapidly circulates hot air forces heat into food faster than would naturally happen due to simple heating without the fan.

### Natural convection

Natural convection, or free convection, occurs due to temperature differences which affect the density, and thus relative buoyancy, of the fluid. Heavier (more dense) components will fall, while lighter (less dense) components rise, leading to bulk fluid movement. Natural convection can only occur, therefore, in a gravitational field. A common example of natural convection is the rise of smoke from a fire. It can be seen in a pot of boiling water in which the hot and less-dense water on the bottom layer moves upwards in plumes, and the cool and denser water near the top of the pot likewise sinks.

Natural convection will be more likely and/or more rapid with a greater variation in density between the two fluids, a larger acceleration due to gravity that drives the convection, and/or a larger distance through the convicting medium. Natural convection will be less likely and/or less rapid with more rapid diffusion (thereby diffusing away the thermal gradient that is causing the convection) and/or a more viscous (sticky) fluid. The onset of natural convection can be determined by the Rayleigh number (Ra).

## 3. Radiation

Thermal radiation is electromagnetic radiation generated by the thermal motion of charged particles in matter. All matter with a temperature greater than absolute zero emits thermal radiation. The mechanism is that bodies with a temperature above absolute zero have atoms or molecules with kinetic energies which are changing, and these changes result in charge-acceleration and/or dipole oscillation of the charges that compose the atoms. This motion of charges produces electromagnetic radiation in the usual way. However, the side spectrum of this radiation reflects the wide spectrum of energies and accelerations of the charges in any piece of matter at even a single temperature.

Examples of thermal radiation include the visible light and infrared light emitted by an incandescent light bulb, the infrared radiation emitted by animals and detectable with an infrared camera, and the cosmic microwave background radiation. Thermal radiation is different from thermal convection and thermal conduction—a person near a raging bonfire feels radiant heating from the fire, even if the surrounding air is very cold.

Sunlight is thermal radiation generated by the hot plasma of the Sun. The Earth also emits thermal radiation, but at a much lower intensity and different spectral distribution (infrared rather than visible) because it is cooler. The Earth's absorption of solar radiation, followed by its outgoing thermal radiation is the two most important processes that determine the temperature and climate of the Earth.

**Source from:**
[1] 张寅平,潘毅群,王馨. 建筑环境与设备工程专业. 第三课 HEAT TRANSFER. 北京:中国建筑工业出版社, 2005.
[2] Roger W. Haines HVAC Systems Design Handbook. chapter 18 heat transfer. Fourth Edition.

## Part 6  Exercises and Practices

## 1. Discussion

(1) Try to talk with your partner about the first law and second law of the thermodynamics.

(2) Is there any difference and relationship between the heat conduction, convection and radiation?

## 2. Translation

(1) Entropy measures the molecular disorder of a system. The more mixed a system, the greater its entropy; conversely, an orderly or unmixed configuration is one of low entropy.

(2) Work is the mechanism that transfers energy across the boundary of systems with differing pressures (or force of any kind), always in the direction of the lower pressure; if the total effect produced in the system can be reduced to the raising of a weight, then nothing but work has crossed the boundary.

(3) When the vapor is at a temperature greater than the saturation temperature, it is superheated vapor. The pressure and temperature of superheated vapor are independent properties, since the temperature can increase while the pressure remains constant. Gases are highly superheated vapors.

(4) However, the simplest of these statements can be formulated by considering a system which executes a cycle. By definition, the system experiences no net change of state. Therefore, there can be no net energy interaction between the system and its environment.

(5) When the boundaries of a system are such that it cannot exchange matter with the surroundings, the system is said to be a closed system. The system, however, may exchange energy in the form of heat or work with the surroundings. The boundaries of a closed system may be rigid or may expand or contract, but the mass of a closed system cannot change.

## 3. Writing

Please write out the thermodynamics progress for a heat engine in about 200 words.

# Lesson 2

# Hydrodynamics and Fluid Machinery

## Part 1  Preliminary

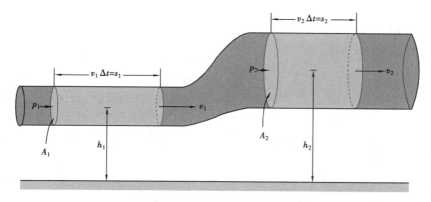

Fig. 2.1  **Bernoulli's principle derivation diagram**

Bernoulli's principle (Fig. 2.1) states that a decrease in the speed of a fluid occurs simultaneously with an increase in pressure or an increase in the fluid's potential energy. Bernoulli's principle can be applied to various types of fluid flow, resulting in various forms of Bernoulli's equation; there are different forms of Bernoulli's equation for different types of flow. In a steady flow, the sum of all forms of energy in a fluid along a streamline is the same at all points on that streamline. This requires that the sum of kinetic energy, potential energy and internal energy remains constant.

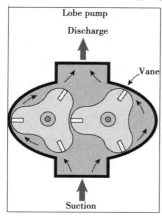

Fig. 2.2  **A common lobe pump**

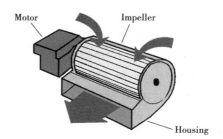

Fig. 2.3  **A typical mechanical fan**

Lobe pumps (Fig. 2.2) are used in a variety of industries including pulp and paper, chemical, food, beverage, pharmaceutical, and biotechnology. They are popular in these diverse industries because they offer superb sanitary qualities, high efficiency, reliability, corrosion resistance and good clean-in-place and steam-in-place (CIP/SIP) characteristics. A mechanical fan (Fig. 2.3) is a powered machine used to create flow within a fluid, typically a gas such as air. A mechanical fan consists of a rotating arrangement of vanes or blades which act on the air. The rotating assembly of blades and hub is known as an impeller, a rotor, or a runner. Usually, it is contained within some form of housing or case. This may direct the airflow or increase safety by preventing objects from contacting the fan blades. Most fans are powered by electric motors, but other sources of power may be used.

## Part 2　Text

流体力学是力学的一个分支,主要研究在各种力的作用下,流体本身的静止状态以及流体和固体界壁间有相对运动时的相互作用和流动规律。本章主要介绍流体的基本属性、流体动力学和流动过程的相关内容,从流体的连续性以及流动过程中的压力损失入手,引入了伯努利方程,并对其进行了深入的阐述和分析,进而介绍了流体的两种基本流动状态:层流和湍流;流动过程主要包括流体与壁面间的摩擦力、边界层理论、流动过程中的分离以及流体的可压缩性等方面的概念和性质。

Based on the continuity of fluid and the pressure loss in the flow process, Bernoulli's equation is introduced and analyzed, and then two basic flow states of fluid are introduced: laminar flow and turbulence flow. The flow process mainly includes the concepts and properties of friction between fluid and wall, boundary layer theory, separation in the flow process and compressibility of fluid. Flowing fluids in heating, ventilating, air-conditioning, and refrigeration systems transfer heat and mass. This chapter introduces the basics of fluid mechanics that are related to HVAC processes, reviews pertinent flow processes, and presents a general discussion of single—phase fluid flow analysis.

### 1. *Fluid properties*

Fluids differ from solids in their reaction to shearing. When placed under shear stress, a solid deforms only a finite amount, whereas a fluid deforms continuously for as long as the shear is applied. Both liquids and gases are fluids. Although liquids and gases differ strongly in the nature of molecular actions, their primary mechanical differences are in the degree of compressibility and liquid formation of a free surface (interface).

Fluid motion can usually be described by one of several simplified modes of action or models. The simplest is the ideal-fluid model, which assumes no resistance to shearing. Ideal flow analysis is well developed, and when properly interpreted is valid for a wide range of applications. Nevertheless, the effects of viscous action may need to be considered. Most fluids in HVAC applications can be treated as Newtonian, where the rate of deformation is directly proportional to the shearing stress. Turbulence complicates fluid behavior, and viscosity influences the nature of the turbulent flow.

## Density

The density ρ of a fluid is its mass per unit volume, The densities of air and water at standard indoor conditions of 20 ℃ and 101.325 kPa are:

$$\rho_{water} = 998 \text{ kg/m}^3$$
$$\rho_{air} = 1.2 \text{ kg/m}^3$$

## Viscosity

Viscosity is the resistance of adjacent fluid layers to shear. For shearing between two parallel plates, each of area $A$ and separated by distance $Y$, the tangential force $F$ per unit area required to slide one plate with velocity $V$, parallel to the other is proportional to $V/Y$:

$$\frac{F}{A} = \mu\left(\frac{V}{Y}\right) \qquad (2.1)$$

Where the proportionality factor $\mu$ is the absolute viscosity or dynamic viscosity of the fluid. The ratio of the tangential force $F$ to area A is the shearing stress $\tau$, and $V/Y$ is the lateral velocity gradient. In complex flows, velocity and shear stress may vary across the flow field; this is expressed by the following differential equation:

$$\tau = \mu \frac{dv}{dy} ① \qquad (2.2)$$

The velocity gradient associated with viscous shear for a simple case involving flow velocity in the $X$ direction but of varying magnitude in the $Y$ direction is distributed as exponential law.

Absolute viscosity $\mu$ depends primarily on temperature. For gases (except near the critical point), viscosity increases with the square root of the absolute temperature, as predicted by the kinetic theory. Liquid viscosity decreases with increasing temperature.

Absolute viscosity has dimensions of force · time/length$^2$. At standard indoor conditions, the absolute viscosities of water and dry air are

$$\mu_{water} = 1.0m \text{ N} \cdot \text{s/m}^2$$
$$\mu_{air} = 18\mu \text{ N} \cdot \text{s/m}^2$$

In fluid dynamics, kinematic viscosity $\nu$ is the ratio of absolute viscosity to density:

$$\nu = \frac{\mu}{\rho} \qquad (2.3)$$

At standard indoor conditions, the kinematic viscosities of water and dry air are:

$$\nu_{water} = 1.00 \times 10^{-6} \text{ m}^2/\text{s}$$
$$\nu_{air} = 16 \times 10^{-6} \text{ m}^2/\text{s}$$

# 2. Basic relations of fluid dynamics

This section considers homogeneous, constant property, incompressible fluids and introduces fluid dynamic considerations used in most analyses.

---

① 见 Part 3 Extension.

## Continuity

Conservation of matter applied to fluid flow in a conduit requires that: $\int \rho v dA = C$

Where,

$v$ = velocity normal to the differential area $dA$;

$\rho$ = fluid density.

Both $\rho$ and $v$ may vary over the cross section $A$ of the conduit. If both $\rho$ and $v$ are constant over the cross-sectional area normal to the flow, then

$$m = \rho VA = C \qquad (2.4)$$

Where $m$ is the mass flow rate across the area normal to the flow. When flow is effectively incompressible, $\rho = C$, in pipeline and duct flow analyses, the average velocity is then $V = (1/A)\int v dA$, The continuity relation is

$$Q = AV = C \qquad (2.5)$$

Where $Q$ is the volumetric flow rate. Except when branches occur, $Q$ is the same at all sections along the conduit.

For the ideal-fluid model, flow patterns around bodies (or in conduit section changes) result from displacement effects. An obstruction in a fluid stream, such as a strut in a flow or a bump on the conduit wall, pushes the flow smoothly out of the way, so that behind the obstruction, the flow becomes uniform again. The effect of fluid inertia (density) appears only in pressure changes.

## Pressure variation across flow

Pressure variation in fluid flow is important and can be easily measured. Variation across streamlines involves fluid rotation (vortices). Lateral pressure variation across streamlines is given by the following relation:

$$\frac{\partial}{\partial r}\left(\frac{p}{\rho} + gz\right) = \frac{v^2}{r} \qquad (2.6)$$

Where,

$r$ = radius of curvature of the streamline;

$z$ = elevation.

This relation explains the pressure difference found between the inside and outside walls of a bend and near other regions of conduit section change. It also states that pressure variation is hydrostatic ($p + \rho gz = C$) across any conduit where streamlines are parallel.

## The <u>Bernoulli Equation</u> [2] and pressure variation along flow

A basic tool of fluid flow analysis is the Bernoulli relation, which involves the principle of energy conservation along a streamline. Generally, the Bernoulli Equation is not applicable across stre-

---

[2] 见 Part 3 Extension.

amlines. The first law of thermodynamics can be applied to mechanical flow energies (kinetic and potential) and thermal energies: heat is a form of energy and energy is conserved.

The change in energy content $\Delta E$ per unit mass of flowing material is a result from the work $W$ done on the system plus the heat $Q$ absorbed:

$$\Delta E = W + Q \tag{2.7}$$

Fluid energy is composed of kinetic, potential, and internal ($\mu$) energies. Per unit mass of fluid, the above energy change relation between two sections of the system is

$$\Delta\left(\frac{v^2}{2} + gz + \mu\right) = E_M - \Delta\left(\frac{p}{\rho}\right) + Q \tag{2.8}$$

Where the work terms are (1) the external work $E_M$ from a fluid machine ($E_M$ is positive for a pump or blower) and (2) the pressure or flow work $p/\rho$. Rearranging, the energy equation can be written as the generalized Bernoulli equation:

$$\Delta\left(\frac{v^2}{2} + gz + \mu\right) + \Delta\mu = E_M + Q \tag{2.9}$$

The term in parentheses in Equation (2.9) is the Bernoulli constant:

$$\frac{p}{\rho} + \frac{v^2}{2} + gz = B \tag{2.10a}$$

In cases with no viscous action and no work interaction, $B$ is constant; more generally its change (or lack thereof) is considered in applying the Bernoulli equation. The terms making up $B$ are fluid energies (pressure, kinetic, and potential) per mass rate of fluid flow. Alternative forms of this relation are obtained through multiplication by $p$ or division by $g$:

$$p + \frac{\rho v^2}{2} + \rho gz = \rho B \tag{2.10b}$$

$$\frac{p}{\rho g} + \frac{v^2}{2g} + z = \frac{B}{g} \tag{2.10c}$$

The first form involves energies per volume flow rate, or pressures; the second involves energies per mass flow rate, or heads. In gas flow analysis, Equation (2.10b) is often used with the $\rho gz$ term dropped as negligible. Equation (2.10a) should be used when density variations occur. For liquid flows, Equation (2.10c) is commonly used. Identical results are obtained with the three forms if the units are consistent and the fluids are homogeneous.

Many systems of pipes or ducts and pumps or blowers can be considered as one-dimensional flow. The Bernoulli equation is then considered as velocity and pressure vary along the conduit. Analysis is adequate in terms of the section-average velocity $V$ of Equation (2.4) or (2.5), In the Bernoulli relation [equations (2.9) and (2.10)], $v$ is replaced by $V$, and variation across streamlines can be ignored; the whole conduit is now taken as one streamline. Two-and three-dimensional details of local flow occurrences are still significant, but their effect is combined and accounted for in factors.

The kinetic energy term of the Bernoulli constant $B$ is expressed as $\alpha V^2/2$, where the kinetic energy factor ($\alpha > 1$) expresses the ratio of the true kinetic energy of the velocity profile to that of the mean flow velocity.

For laminar flow in a wide rectangular channel, $\alpha = 1.54$, and for laminar flow in a pipe, $\alpha = 2.0$. For turbulent flow in a duct $\alpha \approx 1$.

Heat transfer $Q$ may often be ignored. The change of mechanical energy into internal energy $\Delta\mu$ may be expressed as $E_L$. Flow analysis involves the change in the Bernoulli constant ($\Delta B = B_2 - B_1$) between stations 1 and 2 along the conduit, and the Bernoulli equation can be expressed as:

$$\left(\frac{p}{\rho} + \alpha\frac{V^2}{2} + gz\right)_1 + E_M = \left(\frac{p}{\rho} + \alpha\frac{V^2}{2} + gz\right)_2 + E_L \qquad (2.11a)$$

Or, dividing by $g$, in the form as:

$$\left(\frac{p}{\rho g} + \alpha\frac{V^2}{2g} + z\right)_1 + H_M = \left(\frac{p}{\rho g} + \alpha\frac{V^2}{2g} + z\right)_2 + H_L \qquad (2.11b)$$

The factors $E_M$ and $E_L$ are defined as positive, where $gH_M = E_M$ represents energy added to the conduit flow by pumps or blowers, and $gH_L = E_L$ represents energy dissipated, that is, converted into heat as mechanically non-recoverable energy. A turbine or fluid motor thus has a negative $H_M$ or $E_M$. For conduit systems with branches involving inflow or outflow, the total energies must be treated, and analysis is in terms of $mB$ and not $B$.

When real-fluid effects of viscosity or turbulence are included, the continuity relation in Equation (2.5) is not changed, but $V$ must be evaluated from the integral of the velocity profile, using time averaged local velocities.

In fluid flow past fixed boundaries, the velocity at the boundary is zero and shear stresses are produced. The equations of motion then become complex and exact solutions are difficult to find, except in simple cases.

## Laminar flow

For steady, fully developed laminar flow in a parallel-walled conduit, the shear stress $\tau$ varies linearly with distance $y$ from the centerline. For a wide rectangular channel,

$$\tau = \frac{y}{b}\tau_w = \mu\frac{dv}{dy} \qquad (2.12)$$

Where,

$\tau_w$ = wall shear stress = $b(dp/ds)$;
$2b$ = wall spacing;
$s$ = flow direction.

Because the velocity is zero at the wall ($y = b$), the integrated result is:

$$v = \left(\frac{b^2 - y^2}{2\mu}\right)\frac{dp}{ds} \qquad (2.13)$$

This is the Poiseuille-flow parabolic velocity profile for a wide rectangular channel. The average velocity $V$ is two-thirds the maximum velocity (at $y = 0$), and the longitudinal pressure drop in terms of conduit flow velocity is

$$\frac{dp}{ds} = -\left(\frac{3\mu V}{b^2}\right) \qquad (2.14)$$

The parabolic velocity profile can be derived for the axisymmetric conduit (pipe) of radius $R$

but with a different constant. The average velocity is then half the maximum, and the pressure drop relation is:

$$\frac{dp}{ds} = -\left(\frac{8\mu V}{R^2}\right) \quad (2.15)$$

## Turbulence

Fluid flows are generally turbulent, involving random perturbations or fluctuations of the flow (velocity and pressure), characterized by an extensive hierarchy of scales or frequencies. Flow disturbances that are not random, but have some degree of periodicity, such as the oscillating vortex trail behind bodies, have been erroneously identified as turbulence. Only flows involving random perturbations without any order or periodicity are turbulent; the velocity in such a flow varies with time or locale of measurement.

Turbulence can be quantified by statistical factors. Thus, the velocity most often used in velocity profiles is the temporal average velocity $v$, and the strength of the turbulence is characterized by the root-mean-square of the instantaneous variation in velocity about this mean. The effects of turbulence cause the fluid to diffuse momentum, heat, and mass very rapidly across the flow.

The Reynolds number $Re$, a dimensionless quantity, gives the relative ratio of inertial to viscous forces:

$$Re = \frac{VL}{\nu} \quad (2.16)$$

Where,

$L$ = characteristic length;

$\nu$ = kinematic viscosity.

In flow through round pipes and tubes, the characteristic length is the diameter $D$. Generally, laminar flow in pipes can be expected if the Reynolds number, which is based on the pipe diameter, is less than about 2,300. Fully turbulent flow exists when $Re_D > 10,000$. Between 2,300 and 10,000, the flow is in a transition state and predictions are unreliable. In other geometries, different criteria for the Reynolds number exist.

## 3. Basic flow processes

### Wall friction

At the boundary of real-fluid flow, the relative tangential velocity at the fluid surface is zero. Sometimes in turbulent flow studies, velocity at the wall may appear finite, implying a fluid slip at the wall. However, this is not the case; the difficulty is in velocity measurement. Zero wall velocity leads to a high shear stress near the wall boundary and a slowing down of adjacent fluid layers. A velocity profile develops near a wall, with the velocity increasing from zero at the wall to an exterior value within a finite lateral distance.

Laminar and turbulent flow differ significantly in their velocity profiles. Turbulent flow profiles are flat compared to the more pointed profiles of laminar flow. Near the wall, velocities of the turbulent profile must drop to zero more rapidly than those of the laminar profile, so the shear stress and friction are much greater in the turbulent flow case. Fully developed conduit flow may be characterized by the pipe factor, which is the ratio of average to maximum velocity. Viscous velocity profiles result in pipe factors of 0.667 and 0.50 for wide rectangular and axisymmetric conduits.

## Boundary layer

In most flows, the friction of a bounding wall on the fluid flow is evidenced by a boundary layer. For around bodies, this layer encompasses all viscous or turbulent actions, causing the velocity in it to vary rapidly from zero at the wall to that of the outer flow at its edge. Boundary layers are generally laminar near the start of their formation but may become turbulent downstream of the transition point. For conduit flows, spacing between adjacent walls is generally small compared with distance in the flow direction. As a result, layers from the walls meet at the centerline to fill the conduit.

A significant boundary-layer occurrence exists in a pipeline or conduit following a well-rounded entrance. Layers grow from the walls until they meet at the center of the pipe. Near the start of the straight conduit, the layer is very thin, so the uniform velocity core outside has a velocity only slightly greater than the average velocity. As the layer grows in thickness, the slower velocity near the wall requires a velocity increase in the uniform core to satisfy continuity. As the flow proceeds, the wall layers grow until they join, after an entrance length $Le$. Application of Bernoulli relation of Equation (2.10) to the core flow indicates a decrease in pressure along the layer. It is showed that although the entrance length is many diameters, the length in which the pressure drop significantly exceeds those for fully developed flow is on the order of 10 diameters for turbulent flow in smooth pipes.

## Flow patterns with separation

In technical applications, flow with separation is common and often accepted if it is too expensive to avoid. Flow separation may be geometric or dynamic. Geometric separation results when a fluid stream passes over a very sharp corner, as with an orifice; the fluid generally leaves the corner irrespective of how much its velocity has been reduced by friction.

For geometric separation in orifice flow, the outer streamlines separate from the sharp corners and, because of fluid inertia, contract to a section smaller than the orifice opening, the vena contracta, with a limiting area of about six-tenths of the orifice opening. After the vena contracta, the fluid stream expands rather slowly through turbulent or laminar interaction with the fluid along its sides. Outside the jet, fluid velocity is small compared to that in the jet. Turbulence helps spread out the jet, increases the losses, and brings the velocity distribution back to a more uniform profile.

Some geometric separations occur at sharp entrance to a conduit, at an inclined plate or damper in a conduit, and at a sudden expansion. For these, a vena contrata can be identified; for sudden expansion, its area is that of the upstream contraction. Ideal-fluid theory, using tree streamlines, provides insight and predicts contraction coefficients for valves, orifices, and vanes. These geometric flow separations are large loss-producing devices. To expand a flow efficiently or to have an entrance with minimum losses, the device should be designed with gradual contours, a diffuser, or a rounded entrance.

Flow devices with gradual contours are subject to separation that is more difficult to predict, because it involves the dynamics of boundary layer growth under an adverse pressure gradient rather than flow over a sharp corner. In a diffuser, which is used to reduce the loss in expansion, it is possible to expand the fluid some distance at a gentle angle without difficulty.

**Compressibility**

All fluids are compressible to some degree; their density depends on the pressure. Steady liquid flow may ordinarily be treated as incompressible and incompressible flow analysis is satisfactory for gases and vapors at velocities below about 20 to 40 m/s, except in long conduits.

For liquids in pipelines, if flow is suddenly stopped, a severe pressure surge or water hammer is produced that travels along the pipe at the speed of sound in the liquid. This pressure surge alternately compresses and decompresses the liquid. For steady gas flows in long conditions, a decrease in pressure along the conduit can reduce the density of the gas significantly enough to cause the velocity to increase. If the conduit is long enough, velocities approaching the speed of sound are possible at the discharge end, and the Mach number (the ratio of the flow velocity to the speed of sound) must be considered.

Some compressible flows occur without heat gain or loss. If there is no friction, the process is reversible as well. Such a reversible adiabatic process is called isentropic, and the relationship is $p/\rho^k = C$, where $k$ is the ratio of specific heats at constant pressure and volume, has a value of 1.4 for air and diatomic gases.

**Keywords**: Bernoulli Equation, Laminar Flow, Turbulent Flow

**Source from**:
[1] 张寅平, 潘毅群, 王馨. 建筑环境与设备工程专业. 第二课 FLUID FLOW. 北京: 中国建筑工业出版社, 2005.
[2] 赵三元, 闫岫峰. 建筑类专业英语——暖通与燃气. 第七课 LAMINAR AND TURBULENT FLOW. 北京: 中国建筑工业出版社, 2002.

## Part 3　Extension

1. Newton Inner Friction Law 牛顿内摩擦定律

牛顿内摩擦定律只适用于一般流体,对特殊流体是不适用的。为此,将在做纯剪切流动时满足牛顿内摩擦定律的流体称为牛顿流体,如水和空气等,均为牛顿流体;而将不满足该定律的称为非牛顿流体,如泥浆、污水、油漆和高分子溶液等。

2. Bernoulli Equation 伯努利方程

伯努利方程在应用上有很大的灵活性:(1)方程是在恒定流前提下进行的,客观上并不存在绝对的恒定流,但多数流动,流速随时间变化缓慢,由此所导致的惯性力较小,方程仍适用。(2)方程不仅适用于不可压流体流动,还适用于压缩性极小的液体流动,也适用于专业上所碰到的大多数气体流动。(3)方程断面可选在恒定流断面或渐变流断面,但在某些问题中,断面流速不大,离心惯性力不显著,或者断面流速项在能量方程中所占比例很小,也允许将断面划在急变流处,近似地求流速和压强。

## Part 4   Words and Expressions

| | |
|---|---|
| shear | n. [力]切变;修剪 |
| amount | n. 数量;总额,总数 |
| deform | v. 使变形;变形 |
| molecular | adj. [化学]分子的;由分子组成的 |
| compressibility | n. 压缩性;压缩系数;压缩率 |
| resistance | n. 阻力;抵抗;抵抗力 |
| interpret | vt. 解释;说明 |
| valid | adj. 有效的;有根据的 |
| viscous | adj. 粘性的;黏的 |
| viscosity | n. [物]粘性,[物]粘度 |
| turbulent | adj. 湍流的 |
| turbulence | n. [流]湍流 |
| parallel | n. 平行线;对比 |
| tangential force | [力]切线力,[力]切向力 |
| proportional | adj. 比例的,成比例的 |
| dynamic viscosity | 动力粘度 |
| kinematic viscosity | 运动粘度 |
| lateral velocity gradient | 侧向速度梯度 |
| magnitude | n. 大小;量级 |
| exponential | adj. 指数的;n. 指数 |
| critical point | 临界点 |
| square root | 平方根 |
| kinetic | adj. [力]运动的 |
| dimension | n. [数]维;尺寸;次元 |
| continuity | n. 连续性 |
| conservation | n. 保存;保持 |
| conduit | n. 导管;沟渠;导水管 |
| differential | adj. 微分的 |
| cross-sectional | adj. 截面的,断面的,剖面的 |

续表

| volumetric | *adj.* [物]体积的；[物]容积的 |
|---|---|
| obstruction | *n.* 障碍；阻碍；妨碍 |
| bump | *vi.* 碰撞，撞击 |
| variation | *n.* 变化 |
| streamline | *n.* 流线；流线型 |
| rotation | *n.* 旋转；循环，轮流 |
| curvature | *n.* 弯曲，[数]曲率 |
| elevation | *n.* 海拔；高程 |
| hydrostatic | *adj.* 流体静力学的；静水力学的 |
| Bernoulli Equation | 伯努利方程 |
| internal energy | 内能 |
| laminar flow | 层流 |
| turbine | *n.* [动力]涡轮；[动力]涡轮机 |
| centerline | *n.* 中心线 |
| parabolic | *adj.* 抛物线的 |
| profile | *n.* 侧面；轮廓；外形；剖面 |
| longitudinal | *adj.* 长度的，纵向的；经线的 |
| axisymmetric | *adj.* 轴对称的 |
| radius | *n.* 半径，半径范围 |
| perturbation | *n.* 摄动；扰乱 |
| fluctuation | *n.* 起伏，波动 |
| hierarchy | *n.* 层级；等级制度 |
| oscillating | *adj.* [物]振荡的 |
| vortex | *n.* [流]涡流；漩涡 |
| statistical | *adj.* 统计的；统计学的 |
| momentum | *n.* 势头；[物]动量；动力；冲力 |
| Reynolds number | 雷诺数 |
| tangential | *adj.* [数]切线的，[数]正切的 |
| friction | *n.* 摩擦，[力]摩擦力 |
| boundary layer | 边界层 |
| separation | *n.* 分离 |
| geometric | *adj.* 几何学的；[数]几何学图形的 |
| orifice | *n.* [机]孔口 |
| Mach number | 马赫数 |
| diatomic | *adj.* 二原子的；二氢氧基的；二价的 |

# Part 5  Reading

## Computational Fluid Dynamics

### 1. *CFD activity*

Computational fluid dynamics (CFD) is concerned with numerical solution of differential equations governing transport of mass, momentum, and energy in moving fluids. CFD activity emerged and gained prominence with availability of computers in the early 1960s. Today, CFD finds extensive usage in basic and applied research, in design of engineering equipment, and in calculation of environmental and geophysical phenomena. Since the early 1970s, commercial software packages (or computer codes) became available, making CFD an important component of engineering practice in industrial and environmental organizations.

For a long time, design of engineering equipment such as heat exchangers, furnaces, cooling towers, internal combustion engines, gas turbine engines, hydraulic pumps and turbines, aircraft bodies, sea-going vessels, and rockets depended on painstakingly generated empirical information. The same was the case with numerous industrial processes such as casting, welding, alloying, mixing, drying, air-conditioning, and spraying, environmental discharging of pollutants, and so on. The empirical information is typically displayed in the form of correlations or tables and nomograms among the main influencing variables. Such information is extensively availed by designers and consultants from handbooks.

The main difficulty with empirical information is that it is applicable only to the limited range of scales of fluid velocity, temperature, time, or length for which it is generated. Thus, to take advantage of economies of scale, for example, when engineers were called upon to design a higher capacity power plant, boiler furnaces, condensers, and turbines of ever higher dimensions had to be designed for which new empirical information had to be generated all over again. The generation of this new information was by no means an easy task. This was because the information applicable to bigger scales had to be, after all, generated via laboratory-scale models. This required establishment of scaling laws to ensure geometric, kinematic, and dynamic similarities between models and the full-scale equipment. This activity required considerable experience as well as ingenuity, for it is not an easy matter to simultaneously maintain the three aforementioned similarities. The activity had to, therefore, be supported by flow-visualization studies and by simple (typically, one-dimensional) analytical solutions to equations governing the phenomenon under consideration. Ultimately, experience permitted judicious compromises. Being very expensive to generate, such information is often of a proprietary kind. In more recent times, of course, scaling difficulties are encountered in the opposite direction. This is because electronic equipment is considerably miniaturized and, in materials processing, for example, the more relevant phenomena occur at micro-scales.

The potential of fundamental laws (in association with some further empirical laws) for generating widely applicable and scale-neutral information has been known almost ever since they were invented nearly 200 years ago. The realization of this potential, however, has been made possible only with the availability of computers. The past five decades have witnessed almost exponential growth in the speed with which arithmetic operations can be performed on a computer.

By way of reminder, we note that the three laws governing transport are the following:
1. The law of conservation of mass (Transport of mass), and
2. Newton's second law of motion (Transport of momentum), and
3. The first law of thermodynamics (Transport of energy).

## 2. Transport equations

The aforementioned laws are applied to an infinitesimally small control volume located in a moving fluid. This application results in partial differential equations of mass, momentum and energy transfer. Using tensor notation, we can state these laws as follows:

Mass Equation:
$$\frac{\partial \rho}{\partial t} + \nabla \cdot (\rho \vec{v}) = 0 \tag{2.17}$$

Momentum Equation:
$$\frac{\partial (\rho \vec{v})}{\partial t} + \nabla \cdot (\rho \vec{v} \vec{v}) = \rho \vec{F} + \nabla \cdot \overleftrightarrow{\tau}^* \tag{2.18}$$

Energy Equation:
$$\frac{\partial (\rho E)}{\partial t} + \nabla \cdot (\rho E \vec{v}) = \rho \vec{F} \cdot \vec{v} - \nabla \cdot \vec{q} + \nabla \cdot (\overleftrightarrow{\tau}^* \cdot \vec{v}) \tag{2.19}$$

From the point of view of further discussion of numerical methods, it is indeed a happy coincidence that the set of equations (Mass Equation, Momentum Equation, and Energy Equation) can be cast as a single equation for a general variable $\phi$. Thus,

$$\frac{\partial (\rho \phi)}{\partial t} + \nabla \cdot (\rho \vec{v} \varphi) = \nabla \cdot (\Gamma \nabla \phi) + S_\phi \tag{2.20}$$

## 3. Main task

It is now appropriate to list the main steps involved in arriving at numerical solutions to the transport equation. To enhance understanding, an example of an idealized combustion chamber of a gas-turbine engine will be considered.

1. Given the flow situation of interest, define the physical (or space) domain of interest.
2. Select transport equations with appropriate diffusion and source laws. Define boundary conditions on segments of the domain boundary for each variable $\phi$ and the fluid properties.
3. Select points (called nodes) within the domain so as to map the domain with a grid. Construct control volumes around each node.
4. Integrate Equation 1.6 over a typical control volume so as to convert the partial differential equation into an algebraic one.
5. Devise a numerical method to solve the set of algebraic equations.

6. Devise a computer program to implement the numerical method on a computer.
7. Interpret the solution.
8. Display of results.

In this section, we would like to show the most basic knowledge but not intended to provide a survey of all numerical methods. The readers interested on skill development, skills required for problem formulation, computer code writing, and interpretation of results could reference monographs and literatures on CFD.

**Source form:**
J. Blazek. Computational Fluid Dynamicd: Principles and Applications. Chapter 3: Principles of Solution of the Governing Equations. Baden-Daettwil: Alstom Power Ltd, 2001.

## Part 6    Exercises and Practices

### 1. *Discussion*

(1) Please talk about the property of the model of fluid dynamic.

(2) How do the velocity gradient and absolute viscosity affect the boundary layer?

### 2. *Translaiton*

(1) Most fluids in HVAC applications can be treated as Newtonian, where the rate of deformation is directly proportional to the shearing stress. Turbulence complicates fluid behavior, and viscosity influences the nature of the turbulent flow.

(2) Generally, the Bernoulli equation is not applicable across streamlines. The first law of thermodynamics can be applied to mechanical flow energies (kinetic and potential) and thermal energies: heat is a form of energy and energy is conserved.

(3) Many systems of pipes or ducts and pumps or blowers can be considered as one-dimensional flow. The Bernoulli equation is then considered as velocity and pressure vary along the conduit.

(4) Turbulence can be quantified by statistical factors. Thus, the velocity most often used in velocity profiles is the temporal average velocity $v$, and the strength of the turbulence is characterized by the root-mean-square of the instantaneous variation in velocity about this mean.

(5) Near the start of the straight conduit, the layer is very thin, so the uniform velocity core outside has a velocity only slightly greater than the average velocity. As the layer grows in thickness, the slower velocity near the wall requires a velocity increase in the uniform core to satisfy continuity.

### 3. *Writing*

Please describe the formation and characteristics of the boundary layer in about 200 words.

# Lesson 3
# Thermal Comfort

## Part 1  Preliminary

Human body can be viewed as a heat engine where food is the input energy. Thermal neutrality is maintained when the heat generated by human metabolism is allowed to dissipate by convection, radiation, conduction as well as evaporation, thus maintaining thermal equilibrium with the surroundings. The main factors that influence thermal comfort are those that determine heat gain and loss, namely metabolic rate, clothing insulation, air temperature, mean radiant temperature, air speed and relative humidity. Psychological parameters, such as individual expectations, also affect thermal comfort. Maintaining this standard of thermal comfort for occupants of buildings or other enclosures is one of the important goals of HVAC design engineers.

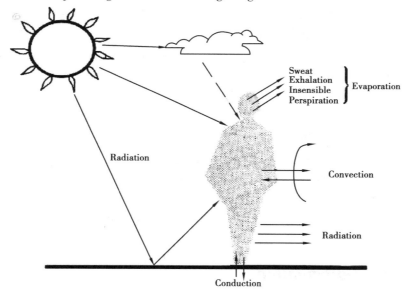

Fig. 3.1  Comfort and the body's heat balance

## Part 2  Text

本章主要介绍建筑物中人员的热舒适性。人体可以看作是一个以食物为输入能量的热机,由新陈代谢产生的热量通过对流、辐射、传导和蒸发等方式不断地释放到周围环境中,从而保持与环境的热平衡、维持机体的正常运转和舒适性。大量的研究表明,影响热舒适的主要因素有人体新陈代谢率、服装热阻、空气温度、平均辐射温度、风速和相对湿度。另外,人们对热湿环境的心理期望也会影响对热舒适的判断。暖通空调设计的重要目的之一,就是维持室内人员的热舒适性。

Building occupants are always in search of thermal comfort, which in turn influences a person's performance (intellectual, manual and perceptual). Depending on the available means, occupants will attempt several actions to change or control environmental conditions. In order to be most successful in these actions, one must have a thorough quantitative, as well as qualitative, knowledge of the conditions establishing the parameters that influence thermal comfort. This will also enable building designers, to provide alternative means to the occupants for controlling their thermal comfort conditions, instead of lowering the thermostat during summer or increasing it during winter.

## 1. *Definition*

A principal purpose of HVAC is to provide conditions for human thermal comfort, "that condition of mind that expresses satisfaction with the thermal environment" (ASHRAE *Standard* 55). This definition leaves open what is meant by "condition of mind" or "satisfaction", but it correctly emphasizes that judgment of comfort is a cognitive process involving many inputs influenced by physical, physiological, psychological, and other processes.

The conscious mind appears to reach conclusions about thermal comfort and discomfort from direct temperature and moisture sensations from the skin, deep body temperatures, and the efforts necessary to regulate body temperatures (Berglund 1995; Gagge 1937; Hardy et al. 1971; Hensel 1973, 1981). In general, comfort occurs when body temperatures are held within narrow ranges, skin moisture is low, and the physiological effort of regulation is minimized.

Comfort also depends on behaviors that are initiated consciously or unconsciously and guided by thermal and moisture sensations to reduce discomfort. Some examples are altering clothing, altering activity, changing posture or location, changing the thermostat setting, opening a window, complaining, or leaving the space.

Surprisingly, although climates, living conditions, and cultures differ widely throughout the world, the temperature that people choose for comfort under similar conditions of clothing, activity, humidity, and air movement has been found to be very similar (Busch 1992; de Dear et al. 1991; Fanger 1972).

## 2. *Human thermoregulation*

Metabolic activities of the body result almost completely in heat that must be continuously dissipated and regulated to maintain normal body temperatures. Insufficient heat loss leads to overheating (hyperthermia), and excessive heat loss results in body cooling (hypothermia). Skin temperature greater than 113 °F or less than 64.5 °F can cause pain (Hardy et al. 1952). Skin temperatures associated with comfort at sedentary activities are 91.5 to 93 °F and decrease with increasing activity (Fanger 1967). In contrast, internal temperatures rise with activity. The temperature regulatory center in the brain is about 98.2 °F at rest in comfort and increases to about 99.3 °F when walking and 100.2 °F when jogging. An internal temperature less than about 82 °F can lead to serious cardiac arrhythmia and death, and a temperature greater than 115 °F can cause irreversible brain damage. Therefore, careful regulation of body temperature is critical to comfort and health.

A resting adult produces about 350 Btu/h of heat. Because most of this is transferred to the environment through the skin, it is often convenient to characterize metabolic activity in terms of heat production per unit area of skin. For a resting person, this is about 18.4 Btu/h · ft$^2$ [50 kcal/(h · m$^2$)] and is called 1 met. This is based on the average male European, with a skin surface area of about 19.4 ft$^2$. For comparison, female Europeans have an average surface area of 17.2 ft$^2$. Systematic differences in this parameter may occur between ethnic and geographical groups. Higher metabolic rates are often described in terms of the resting rate. Thus, a person working at metabolic rate five times the resting rate would have a metabolic rate of 5 met.

The hypothalamus, located in the brain, is the central control organ for body temperature. It has hot and cold temperature sensors and is bathed by arterial blood. Because the recirculation rate of blood is rapid and returning blood is mixed together in the heart before returning to the body, arterial blood is indicative of the average internal body temperature. The hypothalamus also receives thermal information from temperature sensors in the skin and perhaps other locations as well (e.g., spinal cord, gut), as summarized by Hensel (1981).

The hypothalamus controls various physiological processes to regulate body temperature. Its control behavior is primarily proportional to deviations from set-point temperatures with some integral and derivative response aspects. The most important and often-used physiological process is regulating blood flow to the skin: when internal temperatures rise above a set point, more blood is directed to the skin. This vasodilation of skin blood vessels can increase skin blood flow by 15 times [from 0.56 L/(h · ft$^2$) at resting comfort to 8.4 L/(h · ft$^2$)] in extreme to carry internal heat to the skin for transfer to the environment. When body temperatures fall below the set point, skin blood flow is reduced (vasoconstricted) to conserve heat. The effect of maximum vasoconstriction is equivalent to the insulating effect of a heavy sweater. At temperatures less than the set point, muscle tension increases to generate additional heat; where muscle groups are opposed, this may increase to visible shivering, which can increase resting heat production to 4.5 met.

At elevated internal temperatures, sweating occurs. This defense mechanism is a powerful way to cool the skin and increase heat loss from the core. The sweating function of the skin and its control is more advanced in humans than in other animals and is increasingly necessary for comfort at metabolic rates above resting level (Fanger 1967). Sweat glands pump perspiration onto the skin surface for evaporation. If conditions are good for evaporation, the skin can remain relatively dry even at high sweat rates with little perception of sweating. At skin conditions less favorable for evaporation, the sweat must spread on the skin around the sweat gland until the sweat-covered area is sufficient to evaporate the sweat coming to the surface. The fraction of the skin that is covered with water to account for the observed total evaporation rate is termed skin wettedness (Gagge 1937).

Humans are quite good at sensing skin moisture from perspiration (Berglund 1994; Berglund and Cunningham 1986), and skin moisture correlates well with warm discomfort and unpleasantness (Winslow et al. 1937). It is rare for a sedentary or slightly active person to be comfortable with skin wettedness greater than 25%. In addition to the perception of skin moisture, skin wettedness increases the friction between skin and fabrics, making clothing feel less pleasant and fabrics feel coarser (Gwosdow et al. 1986). This also occurs with architectural materials and surfaces, particularly smooth, nonhygroscopic surfaces.

With repeated intermittent heat exposure, the set point for the onset of sweating decreases and the proportional gain or temperature sensitivity of the sweating system increases (Gonzalez et al. 1978; Hensel 1981). However, under long-term exposure to hot conditions, the set point increases, perhaps to reduce the physiological effort of sweating. Perspiration as secreted has a lower salt concentration than interstitial body fluid or blood plasma. After prolonged heat exposure, sweat glands further reduce the salt concentration of sweat to conserve salt.

At the surface, the water in sweat evaporates while the dissolved salt and other constituents remain and accumulate. Because salt lowers the vapor pressure of water and thereby impedes its evaporation, the accumulating salt results in increased skin wettedness. Some of the relief and pleasure of washing after a warm day is related to the restoration of a hypotonic sweat film and decreased skin wettedness. Other adaptations to heat are increased blood flow and sweating in peripheral regions where heat transfer is better. Such adaptations are examples of integral control.

## 3. *Energy Balance*[①]

Fig. 3.2 shows the thermal interaction of the human body with its environment. The total metabolic rate M within the body is the metabolic rate required for the person's activity $M_{act}$ plus the metabolic level required for shivering $M_{shiv}$ (should shivering occur). A portion of the body's energy production may be expended as external work $W$; the net heat production $M-W$ is transferred to the environment through the skin surface ($q_{sk}$) and respiratory tract ($q_{res}$) with any surplus or deficit stored ($S$), causing the body's temperature to rise or fall.

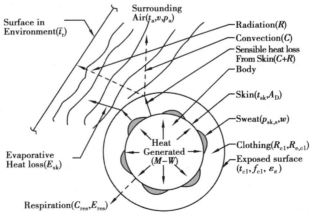

Fig. 3.2 **Thermal interaction of human body and environment**

---

① 见 Part 3 Extension.

$$M - W = q_{sk} + q_{res} + S = (C + R + E_{sk}) + (C_{res} + E_{res}) + (S_{sk} + S_{cr}) \quad (3.1)$$

Where,

$M$ = rate of metabolic heat production, Btu/(h · ft²);

$W$ = rate of mechanical work accomplished, Btu/(h · ft²);

$q_{sk}$ = total rate of heat loss from skin, Btu/(h · ft)²;

$q_{res}$ = total rate of heat loss through respiration, Btu/(h · ft²);

$C + R$ = sensible heat loss from skin, Btu/(h · ft)²;

$E_{sk}$ = total rate of evaporative heat loss from skin, Btu/(h · ft²);

$C_{res}$ = rate of convective heat loss from respiration, Btu/(h · ft²);

$E_{res}$ = rate of evaporative heat loss from respiration, Btu/(h · ft²);

$S_{sk}$ = rate of heat storage in skin compartment, Btu/(h · ft²);

$S_{cr}$ = rate of heat storage in core compartment, Btu/(h · ft²).

Heat dissipates from the body to the immediate surroundings by several modes of heat exchange: sensible heat flow $C + R$ from the skin; latent heat flow from sweat evaporation $E_{rsw}$ and from evaporation of moisture diffused through the skin $E_{dif}$; sensible heat flow during respiration $C_{res}$; and latent heat flow from evaporation of moisture during respiration $E_{res}$. Sensible heat flow from the skin may be a complex mixture of conduction, convection, and radiation for a clothed person; however, it is equal to the sum of the convection $C$ and radiation $R$ heat transfer at the outer clothing surface (or exposed skin).

Sensible and latent heat losses from the skin are typically expressed in terms of environmental factors, skin temperature $t_{sk}$, and skin wettness $w$. Factors also account for the thermal insulation and moisture permeability of clothing. The independent environmental variables can be summarized as air temperature $t_a$, means radiant temperature, relative air velocity $V$, and ambient water vapor pressure $p_a$. The independent personal variables that influence thermal comfort are activity and clothing.

## 4. Conditions for thermal comfort

In addition to the independent environmental and personal variables influencing thermal response and comfort, other factors may also have some effect. These secondary factors include non-uniformity of the environment, visual stimuli, age, and outdoor climate. Studies by Rohles (1973) and Rohles and Nevins (1971) on 1600 college-age students revealed correlations between comfort level, temperature, humidity, sex, and length of exposure. The thermal sensation scale developed for these studies is called the ASHRAE thermal sensation scale:

+ 3      hot

+ 2      warm

+ 1      slightly warm

  0      neutral

- 1    slightly cool
- 2    cool
- 3    cold

Current and past studies are periodically reviewed to update ASHRAE Standard 55, which specifies conditions or comfort zones where 80% of sedentary or slightly active persons find the environment thermally acceptable.

Because people wear different levels of clothing depending on the situation and seasonal weather, ASHRAE Standard 55-2004 defines comfort zones for 0.5 and 1.0 clo [0.44 and 0.88 (ft² · h · °F)/Btu] clothing levels (Fig. 3.3). For reference, a winter business suit has about 1.0 clo of insulation, and a short-sleeved shirt and trousers has about 0.5 clo. The warmer and cooler temperature borders of the comfort zones are affected by humidity and coincide with lines of constant **ET**\*②. In the middle of a zone, a typical person wearing the prescribed clothing would have a thermal sensation at or very near neutral. Near the boundary of the warmer zone, a person would feel about +0.5 warmer on the ASHRAE thermal sensation scale; near the boundary of the cooler zone, that person may have a thermal sensation of -0.5.

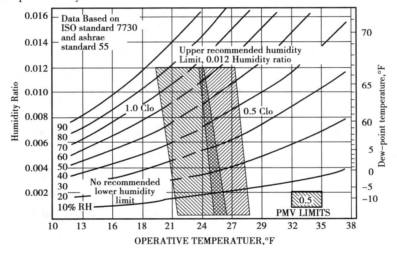

**Fig. 3.3  ASHRAE summer and winter comfort zones**

[Acceptable ranges of operative temperature and humidity with air speed ≤ 0.2 m/s for people wearing 1.0 and 0.5 clo clothing during primarily sedentary activity (≤1.1 met)]

The comfort zone's temperature boundaries ($T_{min}$, $T_{max}$) can be adjusted by interpolation for clothing insulation levels ($I_{cl}$) between those in Fig. 3.3 by using the following equations:

$$T_{min,I_{cl}} = \frac{(I_{cl} - 0.5\text{clo})T_{min,1.0\text{clo}} + (1.0\text{clo} - I_{cl})T_{min,0.5\text{clo}}}{0.5\text{clo}} \quad (3.2)$$

$$T_{max,I_{cl}} = \frac{(I_{cl} - 0.5\text{clo})T_{max,1.0\text{clo}} + (1.0\text{clo} - I_{cl})T_{max,0.5\text{clo}}}{0.5\text{clo}} \quad (3.3)$$

---

② 见 Part 3 Extension.

In general, comfort temperatures for other clothing levels can be approximated by decreasing the temperature borders of the zone by 1°F for each 0.1 clo increase in clothing insulation and vice versa. Similarly, a zone's temperatures can be decreased by 2.5°F per met increase in activity above 1.2 met.

**Keywords**: Thermal comfort, Human thermoregulation, Conditions for thermal comfort

**Source from**:
ASHRAE. 2009 ASHRAE Handbook-Fundamentals. Chapter 9 Thermal Comfort. Alata, GA: ASHRAE, 2009.

## Part 3　Extension

1. Energy Balance（人体热平衡方程）

人体的热舒适状态主要建立在人体与周围环境的热交换的基础之上，即人体新陈代谢的产热量和人体向周围环境的散热量之间的平衡关系。如果要使人体的正常体温保持在恒定范围之内，就必须使人体与环境之间的产热与散热平衡。如图3.2所示，采用一个多层的圆柱断面表示人体的核心部分、皮肤与衣着，人体与环境之间达到热平衡时，需满足热平衡方程式（3.1）。人体热平衡方程作为学科基础，广泛运用在航天服的开发研制、服装的散热性研究、生物医学、矿井开采研究等领域。

2. ET*（新有效温度）

ET*是新有效温度（New Effective Temperature）的简称，是以皮肤湿度变化为基础，反映环境的干球温度、平均辐射温度、湿度对人体热交换的综合作用的热舒适指标。人在不同湿度的实验环境中达到热平衡以后，与相对湿度为50%的均匀温度空间的辐射、对流、蒸发的换热量相同时，均匀空气的温度值为新有效温度值。

## Part 4　Words and Expressions

| | |
|---|---|
| thermal comfort | 热舒适 |
| discomfort | n. 不安，不舒适，不舒服； |
| | vt. 使……不舒服；使……不安 |
| thermostat | n. 恒温(调节)器，自动调温器 |
| thermal environment | 热环境 |
| body temperature | 人体温度 |
| internal temperature | 核心温度 |
| skin temperature | 皮肤温度 |
| thermal sensation | 热感觉 |
| moisture sensation | 湿度感 |
| heat gain | 得热 |
| heat loss | 散热 |
| metabolic activity | 新陈代谢活动 |
| metabolic rate | 新陈代谢率 |

续表

| at sedentary | 静坐 |
| --- | --- |
| temperature sensor | 温度传感器 |
| skin wettedness | 皮肤湿润度（与 skin moisture 同义） |
| perspiration | *n.* 汗，汗水，出汗，流汗 |
| heat exposure | 热暴露 |
| impede | *vt.* 阻碍，妨碍，阻止 |
| adaptation | *n.* 适应，顺应；适应性的改变；改编，改编本 |
| energy balance | 能量平衡 |
| sensible heat | 显热 |
| latent heat | 潜热 |
| thermal sensation scale | 热感觉标度 |

# Part 5　Reading

## Thermal Nonuniform Conditions and Local Discomfort

　　A person may feel thermally neutral as a whole but still feel uncomfortable if one or more parts of the body are too warm or too cold. Nonuniformities may be due to a cold window, a hot surface, a draft, or a temporal variation of these. Even small variations in heat flow cause the thermal regulatory system to compensate, thus increasing the physiological effort of maintaining body temperatures. The boundaries of the comfort zones of ASHRAE Standard 55 provide a thermal acceptability level of 90% if the environment is thermally uniform. Because the standard's objective is to specify conditions for 80% acceptability, the standard permits nonuniformities to decrease acceptability by 10%. Fortunately for the designer and user, the effect of common thermal nonuniformities on comfort is quantifiable and predictable, as discussed in the following sections. Furthermore, most humans are fairly insensitive to small nonuniformities.

### 1. *Asymmetric thermal radiation*

　　Asymmetric or **nonuniform thermal radiation**（不对称热辐射）in a space may be caused by cold windows, uninsulated walls, cold products, cold or warm machinery, or improperly sized heating panels on the wall or ceiling. In residential buildings, offices, restaurants, etc., the most common causes are cold windows or improperly sized or installed ceiling heating panels. At industrial workplaces, the reasons include cold or warm products, cold or warm equipment, etc.

　　Recommendations in ISO Standard 7730 and ASHRAE Standard 55 are based primarily on studies reported by Fanger et al. (1980). These standards include guidelines regarding the radiant temperature asymmetry from an overhead warm surface (heated ceiling) and a vertical cold surface (cold window). Among the studies conducted on the influence of asymmetric thermal radiation are those by Fanger and Langkilde (1975), McIntyre (1974, 1976), McIntyre and Griffiths (1975), McNall and Biddison (1970), and Olesen et al. (1972). These studies all used seated subjects, who were always in thermal neutrality and exposed only to the discomfort resulting from excessive asymmetry.

The subjects gave their reactions on their comfort sensation, and a relationship between the radiant temperature asymmetry and the number of subjects feeling dissatisfied was established (Fig. 3.4). Radiant asymmetry, as defined in the section on Environmental Parameters, is the difference in radiant temperature of the environment on opposite sides of the person. More precisely, radiant asymmetry is the difference in radiant temperatures seen by a small flat element looking in opposite directions.

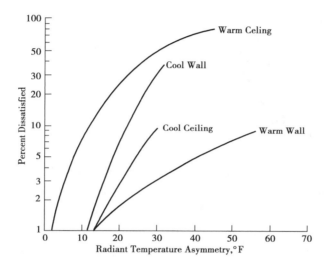

Fig. 3.4  **Percentage of people expressing discomfort due to asymmetric radiation**

Fig. 3.4 shows that people are more sensitive to asymmetry caused by an overhead warm surface than by a vertical cold surface. The influence of an overhead cold surface or a vertical warm surface is much less. These data are particularly important when using radiant panels to provide comfort in spaces with large cold surfaces or cold windows.

Other studies of clothed persons in neutral environments found thermal acceptability unaffected by radiant temperature asymmetries of 10 K or less (Berglund and Fobelets 1987) and comfort unaffected by asymmetries of 20 K or less (McIntyre and Griffiths 1975).

## 2. *Draft*

Draft is an undesired local cooling of the human body caused by air movement. This is a serious problem, not only in many ventilated buildings but also in automobiles, trains, and aircraft. Draft has been identified as one of the most annoying factors in offices. When people sense draft, they often demand higher air temperatures in the room or that ventilation systems be stopped.

Fanger and Christensen (1986) aimed to establish the percentage of the population feeling draft when exposed to a given mean velocity. Fig. 3.5 shows the percentage of subjects who felt draft on the head region (the dissatisfied) as a function of mean air velocity at the neck. The head region comprises head, neck, shoulders, and back. Air temperature significantly influenced the percentage of dissatisfied. There was no significant difference between responses of men and women. The data in Fig. 3.5 apply only to persons wearing normal indoor clothing and performing light, mainly sedentary work. Persons with higher activity levels are not as sensitive to draft (Jones et al. 1986).

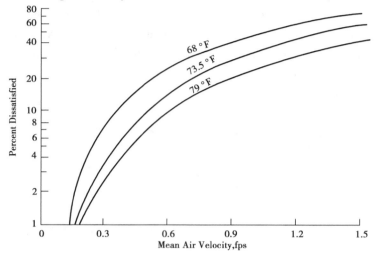

**Fig. 3.5 Percentage of people dissatisfied as function of mean air velocity**

A study of the effect of air velocity over the whole body found thermal acceptability unaffected in neutral environments by air speeds of 0.25 m/s or less (Berglund and Fobelets 1987). This study also found no interaction between air speed and radiant temperature asymmetry on subjective responses. Thus, acceptability changes and the percent dissatisfied because of draft and radiant asymmetry are independent and additive.

Fanger et al. (1989) investigated the effect of turbulence intensity on sensation of draft. Turbulence intensity significantly affects draft sensation, as predicted by the following model. This model can be used to quantify draft risk in spaces and to develop air distribution systems with a low draft risk.

$$PD = (34 - t_a)(V - 0.05)^{0.62}(0.37V \times Tu + 3.14) \tag{3.4}$$

Where $PD$ is percent dissatisfied and $Tu$ is the turbulence intensity in % defined by:

$$Tu = 100 \frac{V_{sd}}{V} \tag{3.5}$$

For $V < 0.05$ m/s, insert $V = 0.05$, and for $PD > 100\%$, insert $PD = 100\%$. $V_{sd}$ is the standard deviation of the velocity measured with an omnidirectional anemometer having a 0.2 s time constant.

The model extends the Fanger and Christensen (1986) draft chart model to include turbulence intensity. In this study, $Tu$ decreases when $V$ increases. Thus, the effects of $V$ for the experimental data to which the model is fitted are $20 < t_a < 26$ ℃, $0.05 < V < 0.5$ m/s, and $0 < Tu < 70\%$. Fig. 3.6 gives more precisely the curves that result from intersections between planes of constant Tu and the surfaces of $PD = 15\%$.

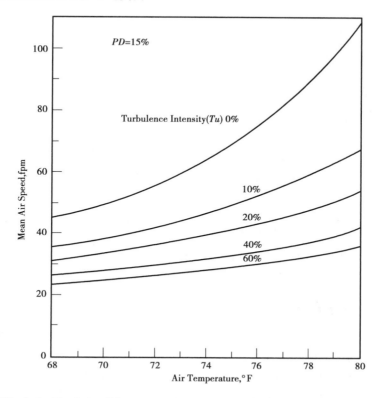

Fig. 3.6  **Draft conditions dissatisfying 15% of population ($PD = 15\%$)**

## 3. *Vertical air temperature difference*

In most buildings, air temperature normally increases with height above the floor. If the gradient is sufficiently large, local warm discomfort can occur at the head and/or cold discomfort can occur at the feet, although the body as a whole is thermally neutral. Among the few studies of vertical air **temperature differences**(垂直温差) and the influence of thermal comfort reported are Eriksson (1975), McNair (1973), McNair and Fishman (1974), and Olesen et al. (1979). Subjects were seated in a climatic chamber so they were individually exposed to different air temperature differences between head and ankles (Olesen et al. 1979). During the tests, the subjects were in thermal neutrality because they were allowed to change the temperature level in the test room whenever they desired; the vertical temperature difference, however, was kept unchanged. Subjects gave subjective reactions to their thermal sensation; Fig. 3.7 shows the percentage of dissatisfied as a function of the vertical air temperature difference between head (1.1 m above the floor) and ankles (0.1 m above the floor).

A head-level air temperature lower than that at ankle level is not as critical for occupants. Eriksson (1975) indicated that subjects could tolerate much greater differences if the head were cooler. This observation is verified in experiments with asymmetric thermal radiation from a cooled ceiling (Fanger et al. 1985).

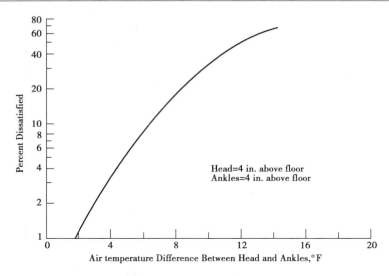

Fig. 3.7 Percentage of seated people dissatisfied as function of air temperature difference between head and ankles

## 4. *Warm or cold floors*

Because of direct contact between the feet and the floor, local discomfort of the feet can often be caused by a too-high or too-low floor temperature. Also, floor temperature significantly influences a room's mean radiant temperature. Floor temperature is greatly affected by building construction (e. g. , insulation of the floor, above a basement, directly on the ground, above another room, use of floor heating, floors in radiant heated areas). If a floor is too cold and the occupants feel cold discomfort in their feet, a common reaction is to increase the temperature level in the room; in the heating season, this also increases energy consumption. A radiant system, which radiates heat from the floor, can also prevent discomfort from cold floors.

The most extensive studies of the influence of floor temperature on feet comfort were performed by Olesen (1977a, 1977b), who, based on his own experiments and reanalysis of the data from Nevins and Feyerherm (1967), Nevins and Flinner (1958), and Nevins et al. (1964), found that flooring material is important for people with bare feet (e. g. , in swimming halls, gymnasiums, dressing rooms, bathrooms, bedrooms). Ranges for some typical floor materials are as follows:

Textiles (rugs)     21 to 28 ℃
Pine floor          22.5 to 28 ℃
Oak floor           24.5 to 28 ℃
Hard linoleum       24 to 28 ℃
Concrete            26 to 28.5 ℃

To save energy, flooring materials with a low contact coefficient (cork, wood, and carpets), radiant heated floors, or floor heating systems can be used to eliminate the desire for higher ambient temperatures caused by cold feet. These recommendations should also be followed in schools, where children often play directly on the floor.

For people wearing normal indoor footwear, flooring material is insignificant. Olesen (1977b) found an optimal temperature of 25 ℃ for sedentary and 23 ℃ for standing or walking persons. At the optimal temperature, 6% of occupants felt warm or cold discomfort in the feet. Fig. 3.8 shows the relationship between floor temperature and percent dissatisfied, combining data from experiments with seated and standing subjects. In all experiments, subjects were in thermal neutrality; thus, the percentage of dissatisfied is only related to discomfort caused by cold or warm feet. No significant difference in preferred floor temperature was found between females and males.

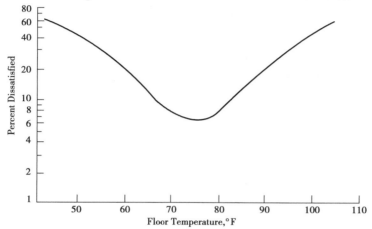

Fig. 3.8  Percentage of people dissatisfied as function of floor temperature

**Source from:**
ASHRAE. 2009 ASHRAE Handbook-Fundamentals. Chapter 9 Thermal Comfort. Alata, GA: ASHRAE, 2009.

## Part 6  Exercise and Practice

### 1. *Translation*

(1) Our thermal interaction with the environment is directed towards maintaining this stability in a process called 'thermoregulation'. The opportunities available to us for such interactions are numerous and complex and are the subject of a great deal of research.

(2) The way in which heat is transferred from our body to the environment is studied by thermal physicists. While sociologists analyze the way we react to the environment. Finally it is the role of the architect or building engineer to design buildings that best meet our thermal needs.

(3) There is a necessary balance between the heat produced by the body and the heat lost from it, if life is to continue. But more is needed to explain the ways in which people achieve thermal comfort.

(4) One can imagine situations in which balance would occur but which might not be considered comfortable. For instance a person with a warm head and cold feet may be in thermal balance but their comfort is not assured! Again, a person who is shivering with cold might be in thermal balance, but is not comfortable.

(5) It is possible to be in thermal balance but still feel too hot or cold. So the determination of comfort conditions is in two stages: first, finding the conditions for thermal balance and then, second, determining which of the conditions so defined are consistent with comfort.

## 2. Discussions

(1) According to your own life experience, please analyze the influence of changing clothes on thermal comfort in the cold season.

(2) Please discuss the factors you think can affect thermal comfort.

## 3. Writing

Please describe how to improve your thermal comfort under a given condition in winter or summer in 200~300 words.

# Lesson 4
# Indoor Environmental Health

## Part 1  Preliminary

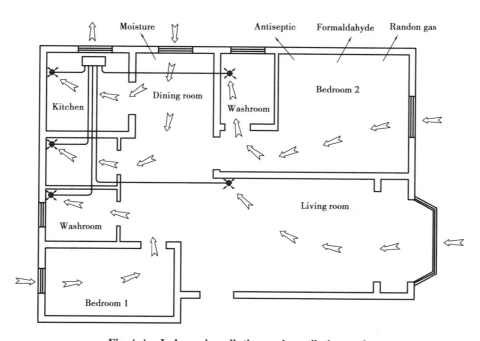

Fig. 4.1  Indoor air pollution and ventilation path

This lesson is about indoor environmental quality (IEQ) and human's health. IEQ includes indoor air quality (IAQ) as well as other physical and psychological aspects of life indoors (e.g., lighting, visual quality, acoustics, and thermal comfort). IAQ is known to affect the health, comfort and well-being of building occupants. Poor indoor air quality has been linked to Sick Building Syndrome (SBS). It can be affected by gases (including carbon monoxide, radon, VOCs), particulates, microbial contaminants (mold, bacteria), or any mass or energy stressor that can induce adverse health conditions. Source control, filtration and the use of ventilation to dilute contaminants are the primary methods for improving indoor air quality in most buildings.

## Part 2  Text

本章主要介绍建筑室内环境品质与人员健康,室内环境品质包括了室内空气品质、声环境、光环境及热舒适等方面。室内空气品质的好坏直接影响着建筑室内人员的身体健康和舒适性。糟糕的室内空气品质可能会引发病态建筑综合征,这有可能是由室内气体污染物(CO、氡气、挥发性有机化合物)、颗粒物、微生物污染(霉菌、细菌)或可能导致不利于人员健康状况的任何物质或能量应激物引起的。对大多数建筑而言,污染源控制、过滤和通风稀释污染物是改善室内空气质量的主要方法。

Indoor environmental health comprises those aspects of human health and disease that are determined by factors in the indoor environment. It also refers to the theory and practice of assessing and controlling factors in the indoor environment that can potentially affect health. The practice of indoor environmental health requires consideration of chemical, biological, physical and ergonomics hazards.

It is essential for engineers to understand the fundamentals of indoor environmental health because the design, operation, and maintenance of buildings and their HVAC systems significantly affect the health of building occupants. In many cases, buildings and systems can be designed and operated to reduce the exposure of occupants to potential hazards. Unfortunately, neglecting to consider indoor environmental health can lead to conditions that create or worsen those hazards.

## 1. **Background**

The most clearly defined area of indoor environmental health is occupational health, particularly as it pertains to workplace airborne contaminants. Evaluation of exposure incidents and laboratory studies with humans and animals have generated reasonable consensus on safe and unsafe workplace exposures for about 1,000 chemicals and particles. Consequently, many countries regulate exposures of workers to these agents. However, chemical and dust contaminant concentrations that meet occupational health criteria usually exceed levels found acceptable to occupants in nonindustrial spaces such as offices, schools, and residences, where exposures often last longer and may involve mixtures of many contaminants and a less robust population (e. g., infants, the elderly, and the infirm) (NAS 1981).

Operational definitions of health, disease, and discomfort are controversial (Cain et al. 1995). However, the most generally accepted definition is that in the constitution of the World Health. Organization (WHO): "Health is a state of complete physical, mental, and social well-being and not merely the absence of disease or infirmity."

Definitions of comfort also vary. Comfort encompasses perception of the environment (e. g., hot/cold, humid/dry, noisy/quiet, bright/dark) and a value rating of affective implications (e. g., too hot, too cold). Rohles et al. (1989) noted that acceptability may represent a more useful concept of evaluating occupant response, because it allows progression toward a concrete goal. Acceptability is the foundation of a number of standards covering thermal comfort and acoustics. Nevertheless, acceptability varies between climatic regions and cultures, and may change over time as expectations change.

Concern about the health effects associated with indoor air dates back several hundred years, and has increased dramatically in recent decades. This attention was partially the result of increased reporting by building occupants of complaints about poor health associated with exposure to indoor air. Since then, two types of diseases associated with exposure to indoor air have been identified: **Sick Building Syndrome**[①](**SBS**) and Building-Related Illness (BRI).

SBS describes a number of adverse health symptoms related to occupancy in a "sick" building, including mucosal irritation, fatigue, headache, and, occasionally, lower respiratory symptoms and nausea. There is no widespread agreement on an operational definition of SBS. Some authors define it as acute discomfort (e.g., eye, nose, or throat irritation; sore throat; headache; fatigue; skin irritation; mild neurotoxic symptoms; nausea; building odors) that persists for more than two weeks at frequencies significantly greater than 20%; with a substantial percentage of complainants reporting almost immediate relief upon exiting the building.

The increased prevalence of health complaints among office workers is typical of sick building syndrome (Burge et al. 1987; Skov and Valbjorn 1987). Widespread occurrence of these symptoms has prompted the World Health Organization to classify SBS into several categories (Morey et al. 1984):

- Sensory irritation in the eyes, nose or throat;
- Skin irritation;
- Neurotoxic symptoms;
- Odor and taste complaints.

Sick building syndrome is characterized by an absence of routine physical signs and clinical laboratory abnormalities. The term nonspecific is sometimes used to imply that the pattern of symptoms reported by afflicted building occupants is not consistent with that for a particular disease. Additional symptoms can include nosebleeds, chest tightness, and fever.

Some investigations have sought to correlate SBS symptoms with reduced neurological and physiological performance. In controlled studies, SBS symptoms can reduce performance in susceptible individuals (Molhave et al. 1986).

Building-related illnesses, in contrast, have a known origin, may have a different set of symptoms, and are often accompanied by physical signs and abnormalities that can be clinically identified with laboratory measurements. For example, hypersensitivity illnesses, including hypersensitivity pneumonitis, humidifier fever, asthma and allergic rhinitis, are caused by individual sensitization to bioaerosols.

## 2. *Descriptions of selected health sciences*

The study of health effects in indoor environments includes a number of scientific disciplines. A few are briefly described here to further the engineer's understanding of which health sciences may be applicable to a given environmental health problem.

---

① 见 Part 3 Extension.

## Epidemiology and biostatistics

Epidemiology studies the cause, distribution, and control of disease in human and animal populations. It represents the application of quantitative methods to evaluate health-related events and effects. Epidemiology is traditionally subdivided into observational and analytical components; the focus may be descriptive, or may attempt to identify causal relationships. Some classical criteria for determining causal relationships in epidemiology are consistency, temporality, plausibility, specificity, strength of association, and dose/response.

Observational epidemiology studies are generally performed with a defined group of interest because of a specific exposure or risk factor. A control group is selected on the basis of similar criteria, but without the exposure or risk factor present. A prospective study (cohort study) consists of observations of a specific group.

Examples of epidemiological investigations are cross-sectional, experimental, and case-control studies. Observations conducted at one point in time are considered cross-sectional studies. In experimental studies, individuals are selectively exposed to a specific agent or condition. These studies are performed with the consent of the participants unless the condition is part of the usual working condition and it is known to be harmless. Control groups must be observed in parallel. Case-control studies are conducted by identifying individuals with the condition of interest and comparing factors of interest in individuals without that condition.

## Industrial hygiene

Industrial hygiene is the science of anticipating, recognizing, evaluating, and controlling workplace conditions that may cause worker illness or injury. Important aspects of industrial hygiene include identifying toxic exposures and physical stressors, determining methods for collecting and analyzing contaminant samples, evaluating measurement results, and developing control measures. Industrial hygienists also create regulatory standards for the work environment, prepare programs to comply with regulations, and collaborate with epidemiologists in studies to document exposures and potential exposures to help determine occupation-related illness.

## Microbiology and mycology

Microbiology studies microorganisms, including bacteria, viruses, fungi, and parasites; mycology is a subspecialty that focuses on fungi. Environmental microbiologists and mycologists investigate the growth, activity, and effect of microorganisms found in nature, many of which can colonize and grow in buildings and building systems. Important aspects of environmental microbiology and mycology are identification of populations of contaminant microorganisms in buildings; determination of methods of collection of air, water, and surface samples; and evaluation of results of microbiological measurements. Effective, practical, and safe disinfection practices are usually developed and validated by microbiologists and mycologists.

Aerobiology is the study of airborne microorganisms or other biologically produced particles, and the effects of these aerosols (bioaerosols) on other living organisms. The section on Bioaerosols has more information on these contaminants.

## Toxicology

Toxicology studies the influence of chemicals on health. All chemical substances may function as toxins, but low concentrations prevent many of them from being harmful. Defining which component of the structure of a chemical predicts the harmful effect is of fundamental importance in toxicology. A second issue is defining the dose/response relationships of a chemical and the exposed population. Dose may refer to delivered dose (exposure presented to the target tissue) or absorbed dose (the dose actually absorbed by the body and available for metabolism). Measures of exposure may be quite distinct from measures of effect because of internal dose modifiers (e.g., delayed metabolism of some toxins because of a lack of enzymes to transform or deactivate them). In addition, the mathematical characteristics of a dose may vary, depending on whether a peak dose, a geometric or arithmetic mean dose, or an integral under the dose curve is used.

Because permission to conduct exposure of human subjects in experimental conditions is difficult to obtain, most toxicological literature is based on animal studies. Isolated animal systems (e.g., homogenized rat livers, purified enzyme systems, or other isolated living tissues) are used to study the effects of chemicals, but extrapolation between dose level effects from animals to humans is problematic.

## 3. *Hazard recognition, analysis and control*

### Hazard recognition and analysis

Hazard recognition and analysis are conducted to determine the presence of hazardous materials or conditions as sources of potential problems. Research, inspection, and analysis determine how a particular hazard affects occupant health. Exposure assessment, an element of hazard recognition, relies on qualitative, semiquantitative, or quantitative approaches. In many situations, air sampling can determine whether a hazardous material is present.

An appropriate sampling strategy must be used to ensure validity of collected samples, determining worst-case (for compliance) or usual (average) exposures. Air sampling can be conducted to determine **Time-Weighted Average**② (**TWA**) exposures, which cover a defined period of time, or short-term exposures, which determine the magnitude of exposures to materials that are acutely hazardous. Samples may be collected for a single substance or a multicomponent mixture. Hazard analysis also characterizes the potential skin absorption or ingestion hazards of an indoor environment. Analyses of bulk material samples and surface wipe samples are also used to determine whether hazardous conditions exist. Physical agent characterization may require direct-reading sampling methods. After collection and analysis, the results must be interpreted and an appropriate control strategy

---

② 见 Part 3 Extension.

developed to control, reduce, or eliminate the hazard.

Hazards are generally grouped into one of the following four classes of environmental stressors:

Chemical hazards. Routes of exposure to airborne chemicals are inhalation (aspiration), dermal (skin) contact, dermal absorption, and ingestion. The degree of risk from exposure depends on the nature and potency of the toxic effects, susceptibility of the person exposed, and magnitude and duration of exposure. Airborne contaminants are very important because of their ease of dispersal from sources and the risk of exposure through the lungs when they are inhaled. Airborne chemical hazards can be gaseous (vapors or gases) or particulate (e.g., dusts, fumes, mists, aerosols, fibers).

Biological hazards. Bacteria, viruses, fungi, and other living or nonliving organisms that can cause acute and chronic illness in workers and building occupants are classified as biological hazards in indoor environments. Routes of exposure are inhalation, dermal (skin) contact, and ingestion. The degree of risk from exposure depends on the nature and potency of the biological hazard, susceptibility of the person exposed, and magnitude and duration of exposure.

Physical hazards. These include excessive levels of ionizing and nonionizing electromagnetic radiation, noise, vibration, illumination, temperature, and force.

Ergonomic hazards. Tasks that involve repetitive motions, require excessive force, or must be carried out in awkward postures can damage muscles, nerves, and joints.

## Hazard control

Strategies for controlling exposures in indoor environment are substitution (removal of the hazardous substance), isolation, disinfection, ventilation, and air cleaning. Not all measures may be applicable to all types of hazards, but all hazards can be controlled by using one of them. Personal protective equipment and engineering, work practice, and administrative controls are used to apply these methods. Source removal or substitution, customarily the most effective measure, is not always feasible. Engineering controls (e.g., ventilation, air cleaning) may be effective for a range of hazards. Local exhaust ventilation is more effective for controlling point-source contaminants than is general dilution ventilation, such as with a building HVAC system.

Hazard Analysis and Control Processes. The goal of hazard analysis and control processes is to prevent harm to people from hazards associated with buildings. Quantitative hazard analysis and control processes are practical and cost-effective. Preventing disease from hazards requires facility managers and owners to answer three simple, site-specific questions:

- What is the hazard?
- How can it be prevented from harming people?
- How can it be verified that the hazard has been prevented from harming people?

Seven principles comprise effective hazard analysis and control:

- Use process flow diagrams to perform systematic hazard analysis;
- Identify critical control points (process steps at which the hazard can be eliminated or prevented from harming people);

- Establish hazard control critical limits at each critical control point;
- Establish a hazard control monitoring plan for critical limits at critical control points;
- Establish hazard control corrective actions for each critical limit;
- Establish procedures to document all activities and results;
- Establish procedures to confirm that the plan (1) actually works under operating conditions (validation), (2) is being implemented properly (verification), and (3) is periodically reassessed.

**Keywords**: Indoor environmental health, Indoor air contaminants, Sick building syndrome

**Source from**:
ASHRAE. 2009 ASHRAE Handbook. Fundamentals-Chapter 10. Atlanta, GA: ASHRAE, 2009.

## Part 3  Extension

1. Sick Building Syndrome (SBS, 病态建筑综合征)

病态建筑综合征也称为建筑物综合征,是发生在建筑物中的一种对人体健康的隐性影响,是由建筑物的运行和维持期间与它的最初设计或规定的运行程序不协调所引起,俗称"空调病"。经常出现的不适症状有:疲乏、头晕、头痛、呼吸不畅、气喘胸闷、咽干喉疼、眼干、鼻塞、流涕、流眼泪、感冒症状、耳鸣等。

2. Time-Weighted Average (TWA, 时间加权平均)

时间加权平均或称为时量平均浓度,简称 TWA。TWA 是一个物理指标,表示对一定时间内化学气体浓度的衡量,是用一个特定时间内与不同浓度的化学制品接触的时间来计算。用这种方式可以计算一天或一周内不同(较高与较低)暴露浓度值平均值。

## Part 4  Words and Expressions

| | |
|---|---|
| ergonomics | *n.* 工效学;人类工程学 |
| occupational health | 职业健康,职业卫生 |
| pertain (to)… | *vi.* 适合;关于,有关;附属,从属 |
| airborne contaminants | 空气污染物 |
| infirm | *adj.* (长期)病弱的;年老体弱的 |
| affective | *adj.* 情感的,表达感情的 |
| concrete | *adj.* 具体的,有形的;混凝土制的 |
| | *n.* 混凝土;具体物;(建)钢筋混凝土 |
| | *vt.* 使凝固;用混凝土修筑 |
| Sick Building Syndrome(SBS) | *n.* 病态建筑综合征 |
| Building-related Illness(BRI) | *n.* 建筑关联病 |
| mucosal irritation | 粘膜刺激 |
| sensory irritation | 感官刺激 |
| skin irritation | 皮肤刺激 |

续表

| susceptible individual | | 易感个体 |
|---|---|---|
| hypersensitivity illnesses | | 过敏性疾病 |
| bioaerosols | n. | 生物气溶胶 |
| Industrial Hygiene | | 工业卫生，劳动卫生 |
| hazard | vt. | 冒险，使遭受危险 |
| | n. | 危险；危害 |
| inhalation | n. | 吸入，吸入剂，吸入物 |
| disinfection | n. | 消毒，灭菌 |
| ventilation | n. | 通风，空气流通 |
| air cleaning | | 空气净化 |

# Part 5   Reading

## Air Contaminants

### 1. *Introduction*

Air contamination is a concern for ventilation engineers when it causes problems for building occupants. Engineers need to understand the vocabulary used by the air sampling and building air cleaning industry. This chapter focuses on the types and levels of air contaminants that might enter ventilation systems or be found as indoor contaminants. Industrial contaminants are included only for special cases.

Air is composed mainly of gases. The major gaseous components of clean, dry air near sea level are approximately 21% oxygen, 78% nitrogen, 1% argon, and 0.04% carbon dioxide. Normal outdoor air contains varying amounts of other materials (permanent atmospheric impurities) from natural processes such as wind erosion, sea spray evaporation, volcanic eruption, and metabolism or decay of organic matter. The concentration of permanent atmospheric impurities varies, but is usually lower than that of anthropogenic air contaminants.

Anthropogenic outdoor air contaminants are many and varied, originating from numerous types of human activity. Electric power generating plants, various modes of transportation, industrial processes, mining and smelting, construction, and agriculture generate large amounts of contaminants. These outdoor air contaminants can also be transmitted to the indoor environment. In addition, the indoor environment can exhibit a wide variety of local contaminants, both natural and anthropogenic.

Contaminants that present particular problems in the indoor environment include allergens (e.g., dust mite or cat antigen), tobacco smoke, radon, and formaldehyde.

Air composition may be changed accidentally or deliberately. In sewers, sewage treatment plants, agricultural silos, sealed storage vaults, tunnels, and mines, the oxygen content of air can become so low that people cannot remain conscious or survive. Concentrations of people in confined spaces (theaters, survival shelters, submarines) require that carbon dioxide given off by normal respiratory functions be removed and replaced with oxygen. Pilots of high-altitude aircraft, breathing at greatly reduced pressure, require systems that increase oxygen concentration. Conversely, for divers working at extreme depths, it is common to increase the percentage of helium (氦) in the atmosphere and reduce nitrogen and sometimes oxygen concentrations.

At atmospheric pressure, oxygen concentrations less than 12% or carbon dioxide concentrations greater than 5% are dangerous, even for short periods. Lesser deviations from normal composition can be hazardous under prolonged exposures.

## 2. *Classes of air contaminants*

Air contaminants are generally classified as either particles or gases. (Particles dispersed in air are also known as **aerosols** (气溶胶). In common usage, the terms aerosol, airborne particle, and particulate contaminant are interchangeable.) The distinction between particles and gases is important when determining removal strategies and equipment. Although the motion of particles is described using the same equations used to describe gas movement, even the smallest of particles (approximately 3 nm) are much larger than individual gas molecules, and have a much greater mass and a much lower diffusion rate. Conversely, particles are typically present in much fewer numbers than even trace levels of contaminant gases.

The **particulate** (悬浮颗粒) class covers a vast range of particle sizes, from dust large enough to be visible to the eye (100 μm) to submicroscopic particles that elude most filters (a few nanometers). Particles may be liquid, solid, or have a solid core surrounded by liquid. They are present in the atmosphere at concentrations ranging from 100 particles/$cm^3$ (mass concentration of a few $\mu g/m^3$) in the cleanest environments to millions per cubic centimeter and several hundred $\mu g/m^3$ in polluted urban environments. The following traditional particulate contaminant classifications arise in various situations, and overlap. They are all still in common use.

• Dusts, fumes, and smokes are mostly solid particulate matter, although smoke often contains liquid particles.

• Mists, fogs, and smogs are mostly suspended liquid particles smaller than those in dusts, fumes, and smokes.

• Bioaerosols include primarily intact and fragmentary viruses, bacteria, fungal spores, and plant and animal allergens; their primary effect is related to their biological origin. Common indoor particulate allergens (dust mite allergen, cat dander, house dust, etc.) and endotoxins are included in the bioaerosol class.

• Particulate contaminants may be defined by their size, such as coarse or fine; visible or invisible; or macroscopic, microscopic, or submicroscopic.

• Particles may be described using terms that relate to their interaction with the human respiratory system, such as inhalable and respirable.

The gaseous class covers chemical contaminants that can exist as free molecules or atoms in air. Molecules and atoms are smaller than particles and may behave differently as a result. This class covers two important subclasses:

• Gases, which are naturally gaseous under ambient indoor or outdoor conditions (i. e., their boiling point is less than ambient temperature at ambient pressure).

• Vapors, which are normally solid or liquid under ambient indoor or outdoor conditions (i. e., their boiling point is greater than ambient temperature at ambient pressure), but which evaporate readily.

Through evaporation, liquids change into vapors and mix with the surrounding atmosphere. Like gases, they are formless fluids that expand to occupy the space or enclosure in which they are confined.

Air contaminants can also be classified according to their sources; properties; or the health, safety, and engineering issues faced by people exposed to them. Any of these can form a convenient classification system because they allow grouping of applicable standards, guidelines, and control strategies. Most such special classes include both particulate and gaseous contaminants.

# 3. *Particulate contaminants*

Airborne particulate contamination ranges from dense clouds of desert dust storms to completely invisible and dilute cleanroom particles. It may be anthropogenic or completely natural. It is often a mixture of many different components from several different sources.

Particles occur in a variety of different shapes, including spherical, irregular, and fibers, which are defined as particles with aspect ratio (length-to-width ratio) greater than 3. In describing particle size ranges, *size* is the diameter of an assumed spherical particle.

## Solid particles

Dusts are solid particles projected into the air by natural forces such as wind, volcanic eruption, or earthquakes, or by mechanical processes such as crushing, grinding, demolition, blasting, drilling, shoveling, screening, and sweeping. Some of these forces produce dusts by reducing larger masses, whereas others disperse materials that have already been reduced. Particles are not considered to be dust unless they are smaller than about 100 $\mu m$. Dusts can be mineral, such as rock, metal, or clay; vegetable, such as grain, flour, wood, cotton, or pollen; or animal, including wool, hair, silk, feathers, and leather. Dust is also used as a catch-all term (house dust, for example) that can have broad meaning.

Fumes are solid particles formed by condensation of vapors of solid materials. Metallic fumes are generated from molten metals and usually occur as oxides because of the highly reactive nature of finely divided matter. Fumes can also be formed by sublimation, distillation, or chemical reaction. Such processes create submicrometre airborne primary particles that may agglomerate into larger particle (1 to 2 $\mu m$) clusters if aged at high concentration.

Bioaerosols are airborne biological materials, including viruses and intact and fragments of bacteria, pollen, fungi, and bacterial and fungal spores. Individual viruses range in size from 0.003 to 0.06 μm, although they usually occur as aggregates and are associated with sputum or saliva and therefore are generally much larger. Most individual bacteria range between 0.4 and 5 μm and may be found singly or as aggregates. Intact individual fungal and bacterial spores are usually 2 to 10 μm, whereas pollen grains are 10 to 100 μm, with many common varieties in the 20 to 40 μm range. The size range of allergens varies widely: the allergenic molecule is very small, but the source of the allergen (mite feces or cat dander) may be quite large.

## Liquid particles

Mists are aggregations of small airborne droplets of materials that are ordinarily liquid at normal temperatures and pressure. They can be formed by atomizing, spraying, mixing, violent chemical reactions, evolution of gas from liquid, or escape as a dissolved gas when pressure is released.

Fogs are clouds of fine airborne droplets, usually formed by condensation of vapor, which remain airborne longer than mists. Fog nozzles are named for their ability to produce extra-fine droplets, as compared with mists from ordinary spray devices. Many droplets in fogs or clouds are microscopic and submicroscopic and serve as a transition stage between larger mists and vapors. The volatile nature of most liquids reduces the size of their airborne droplets from the mist to the fog range and eventually to the vapor phase, until the air becomes saturated with that liquid. If solid material is suspended or dissolved in the liquid droplet, it remains in the air as particulate contamination. For example, sea spray evaporates fairly rapidly, generating a large number of fine salt particles that remain suspended in the atmosphere.

## Complex particles

Smokes are small solid and/or liquid particles produced by incomplete combustion of organic substances such as tobacco, wood, coal, oil, and other carbonaceous materials. The term smoke is applied to a mixture of solid, liquid, and gaseous products, although technical literature distinguishes between such components as soot or carbon particles, fly ash, cinders, tarry matter, unburned gases, and gaseous combustion products. Smoke particles vary in size, the smallest being much less than 1 μm in diameter. The average is often in the range of 0.1 to 0.3 μm.

Environmental Tobacco Smoke (ETS) consists of a suspension of 0.01 to 1.0 μm (mass median diameter of 0.3 μm) solid and liquid particles that form as the superheated vapors leaving burning tobacco condense, agglomerate into larger particles, and age. Numerous gaseous contaminants are also produced, including carbon monoxide.

Smog commonly refers to air pollution; it implies an airborne mixture of smoke particles, mists, and fog droplets of such concentration and composition as to impair visibility, in addition to being irritating or harmful. The composition varies among different locations and at different times. The term is often applied to haze caused by a sunlight-induced photochemical reaction involving materials in automobile exhausts. Smog is often associated with temperature inversions in the atmosphere that prevent normal dispersion of contaminants.

## Sizes of airborne particles

Particle size can be defined in several different ways. These depend, for example, on the source or method of generation, visibility, effects, or measurement instrument. Ambient atmospheric particulate contamination is classified by aerosol scientists and the Environmental Protection Agency (**EPA**,美国环保署) by source mode, with common usage now recognizing two primary modes: coarse and fine.

Coarse-mode aerosol particles are largest, and are generally formed by mechanical breaking up of solids. They generally have a minimum size of 1 to 3 μm (EPA 2004). Coarse particles also include bioaerosols such as mold spores, pollen, animal dander, and dust mite particles that can affect the immune system. Coarse-mode particles are predominantly primary, natural, and chemically inert. Road dust is a good example. Chemically, coarse particles tend to contain crustal material components such as silicon compounds, iron, aluminum, sea salt, and vegetative particles.

Fine-mode particles are generally secondary particles formed from chemical reactions or condensing gases. They have a maximum size of about 1 to 3 μm. Fine particles are usually more chemically complex than coarse-mode particles and result from human activity. Smoke is a good example. Chemically, fine aerosols typically include sulfates, organics, ammonium, nitrates, carbon, lead, and some trace constituents. The modes overlap, and their definitions are not precise. In addition, some aerosol researchers recognize additional modes. Fig. 4.2 shows a typical urban distribution, including the chemical species present in each mode.

Recently, there has been increased interest in even smaller particles, known as ultrafine-mode particles, or **nanoparticles** (纳米粒子). Ultrafines have a maximum size of 0.1 μm (100 nm) (EPA 2004). The U.S. National Nanotechnology Initiative (NNI 2008) also uses this definition for nanoparticles.

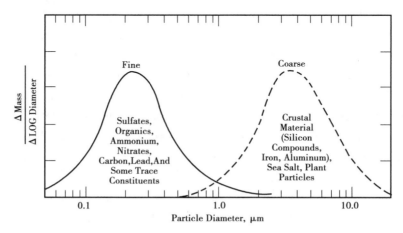

Fig. 4.2 **Typical urban aerosol composition by particle size fraction**
(**EPA** 1982; **Willeke and Baron** 1993)

The size of a particle determines where in the human respiratory system particles are deposited, and various samplers collect particles that penetrate more or less deeply into the lungs. Fig. 4.3 illustrates the relative deposition efficiencies of various sizes of particles in the human nasal and respiratory systems. The inhalable mass is made up of particles that may deposit anywhere in the respiratory system, and is represented by a sample with a median cut point of 100 μm. Most of the inhalable mass is captured in the nasal passages. The thoracic particle mass is the fraction that can penetrate to the lung airways and is represented by a sample with a median cut point of 10 μm (PM10). The respirable particle mass is the fraction that can penetrate to the gas-exchange region of the lungs, which ACGIH (1989) defines as having a median cut point of 4 μm. The EPA no longer uses the term respirable. Their current concern is with particles having a median cut point of 2.5 μm (PM2.5), which they refer to as fine particles.

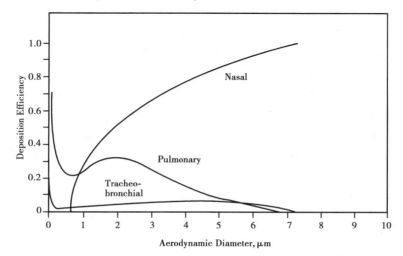

Fig. 4.3 Relative deposition efficiencies of different-sized particles in the three main regions of the human respiratory system, calculated for moderate activity level
(Task Group on Lung Dynamics 1966)

Particles differ in density, and may be irregular in shape. It is useful to characterize mixed aerosol size in terms of some standard particle. The aerodynamic (equivalent) diameter [气体动力学(当量)直径] of a particle, defined as the diameter of a unit-density sphere having the same gravitational settling velocity as the particle in question (Willeke and Baron 1993), is commonly used as the standard particle size. Samplers that fractionate particles based on their inertial properties, such as impactors and cyclones, naturally produce results as functions of the aerodynamic diameters. Samplers that use other sizing principles, such as optical particle counters, must be calibrated to give aerodynamic diameter.

**Source from:**
ASHRAE. 2009 ASHRAE Handbook-Fundamentals. Chapter 11 Air Contaminants. Atlanta, GA: ASHRAE, 2009.

## Part 6  Exercise and Practice

## 1. *Translation*

(1) Often the remedies to problems in existing buildings can be quite drastic. Hazardous materials, buried inside the fabric of buildings, are hard to remove, and eliminating emissions from them can be a major task.

(2) Indoor levels of pollutants may be two to five times higher than the levels of pollutants found outdoors, and this figure can sometimes rise to more than 100 times higher than outdoor levels within schools and other large facilities. In many of today's schools, students are crowded closely together, with class sizes being far larger than they were just a few decades ago.

(3) The main way to create healthier buildings and to reduce hazardous emissions and bad IAQ is to follow a number of simple design principles, which includes the precautionary principle.

(4) Building healthier buildings is not difficult, once architects and their clients accept that buildings do not need to be full of synthetic and hazardous chemicals. If natural and non-toxic materials are used instead, then hazardous emissions can be reduced significantly.

(5) Dependence on ducted and heat recovery mechanical ventilation systems should be avoided at all costs as this will only become a maintenance headache and may add to indoor air problems.

(6) Moisture and dampness can be managed through natural and passive ventilation and the use of breathable and hygroscopic materials.

## 2. *Discussions*

(1) Have you ever suffered from Sick Building Syndrome? If yes, share your exprience.

(2) Do you know how to prevent the occurrence of Sick Building Syndrome? List some means your know.

## 3. *Writing*

Please analyze the ways to improve Indoor Environmental Quality in 200~300 words.

# Lesson 5

# Introduction of Heating System

## Part 1  Preliminary

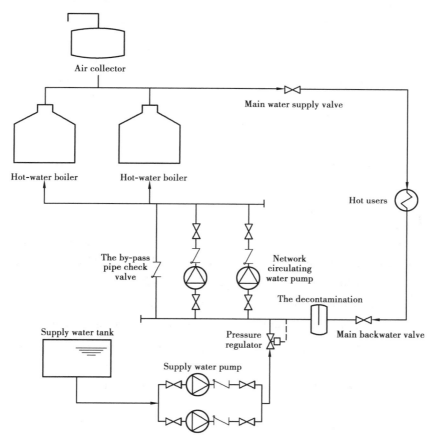

Fig. 5.1  **Hot water heating system**

Water, steam, and air are common heat media for heating systems. The heating system using hot water as the heat medium is called the hot water heating system. It is generally composed of the hot water boiler, the radiator, the water supply pipe, the water return pipe and the expansion water tank. The supply water in the hot water heating system is heated to the required temperature in the boiler, and along the water supply pipe to each hot user forced by the circulating water pump. After releasing the heat, the circulating hot water returns to the boiler along the return pipe.

## Part 2　Text

供暖系统的设计应该根据建筑规模、建筑朝向、建筑功能、人员热舒适要求等,以及所在地区气象条件、能源政策、环保要求等,通过技术经济比较,确定采用哪种方式和类型。设计首先要确定需求(是否考虑人员热舒适、是否需要独立控制、对能耗是否有特定要求等),而后判断设计参数是否满足热舒适需求以及能源限定要求。在节能方面可依据实际条件采取相应措施,比如利用排风与新风换热,实现新风热回收;合理的系统配置,保证系统在不同负荷情况下较高的效率。

This part, which deals with strategic choices, is relatively broad ranging and discursive and is intended to be read from time to time as a reminder of the key decisions to be taken at the start of the design process.

In the design, the designer should firstly fully map the design process for each application will be unique, but will follow the general format:

—problem definition

—ideas generation

—analysis

—selection of the final solution.

This procedure is illustrated in Fig. 5.2 in the form of an outline flowchart.

## 1. General

In common with some other aspects of building services, the requirements placed upon the heating system depend crucially on the form and fabric of the building. It follows that the role of the building services engineer in heating system design is at its greatest when it begins at an early stage, when decisions about the fabric of the building can still be influenced. This allows option for heating to be assessed on an integrated basis that takes account of how the demand for heating is affected by building design as well as by the provision of heating. In other cases, especially in designing replacement heating system for existing building, the scope for integrated design may be much more limited. In all cases, however, the designer should seek to optimize the overall design as far as is possible within the brief.

A successful heating system design will result in a system that can be installed and commissioned to deliver the indoor temperatures required by the client. When in operation, it should operate with high efficiency to minimize fuel costs and environmental emissions while meeting those requirements. It should also sustain its performance over its planned life with limited need for maintenance and replacement of components. Beyond operational and economic requirements, the designer must comply with legal requirements, including those relating to environmental impact and to health and safety.

## 2. Purposes of space heating systems

Heating systems in most buildings are principally required to maintain comfortable conditions for

## Lesson5  Introduction of Heating System

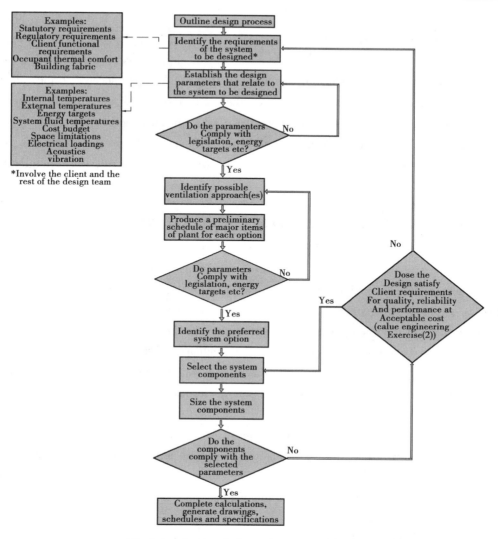

Fig. 5.2  Outline design process: heating

people working or living in the building. As the human body exchanges heat with its surroundings both by convection and by radiation, comfort depends on the temperature of both the air and the exposed surfaces surrounding it and on air movement. **Dry resultant temperature**[①], which combines air temperature and mean radiant temperature, has generally been used for assessing comfort. The predicted mean vote (PMV) index, incorporates a range of factors contributing to thermal comfort.

In buildings (or parts of buildings) that are not normally occupied by people, heating may not be required to maintain comfort. However, it may be necessary to control temperature or humidity in order to protect the fabric of the building or its contents, e. g., from frost or condensation or for processes carried out within the building. In either case, the specific requirements for each room or zone need to be established.

---

① 见 Part 3 Extension.

## 3. Site-related issues

The particular characteristics of the site need to be taken into account, including exposure, site access and connection to gas or heating mains. Exposure is taken into account in the calculation of heat loss. The availability of mains gas or heat supplies is a key factor affecting the choice of fuel.

The form and orientation of buildings can have a significant effect on demand for heating and cooling. If the building services designer is involved early enough in the design process, it will be possible to influence strategic decisions, e.g. to optimize the "passive solar" contribution to energy requirements.

## 4. Legal, economic and general considerations

Various stands of legislation affect design of heating system. Aspects of the design and performance of heating systems are covered by building regulations aimed at the conservation of fuel and power; and regulations set minimum efficiency levels for boilers. Heat producing appliances are also subject to regulations governing supply of combustion air, fuels and chimneys, and emissions of gases and particles to the atmosphere. Designers should also be aware of their obligations to comply with the construction (design and management) regulations and the health and safety at work act.

Beyond strictly legal requirements, the client may wish to meet energy and environmental targets, which can depend strongly on heating system performance. These include:

- Setting energy targets.
- Carbon performance rating/carbon intensity: although primarily intended as a means of showing compliance with the building regulations, "carbon performance rating" (CPR) and "carbon intensity" may be used more widely to define performance. CPR applies to the overall energy performance of office building with air conditioning and mechanical ventilation. Carbon intensity applies to heating systems generally.
- Broader ranging environmental assessments also take energy use into account.
- Clients who own and manage social housing may also have "affordable warmth" targets. Which aim to ensure that low income households will not find their homes too expensive to heat.

Economic appraisal of different levels of insulation, heating systems, fuels, controls should be undertaken to show optimum levels of investment according to the client's own criteria, which may be based on a simple payback period, or a specified discount rate over a given lifetime. Public sector procurement policies may specifically require life cycle costing.

## 5. Interaction with building design, building fabric, services and facilities

As noted above, the earlier the heating system designer can be involved in the overall design process, the greater the scope for optimization. The layout of the building, the size and orientation of windows, the extent and location of thermal mass within the building, and the levels of insulation of the building fabric can all have a significant effect on demand for heat. The airtightness of the building shell and the way in which the building is ventilated are also important. Buildings that are very well insulated and airtight may have no net heating demand when occupied, which requires heating systems to be designed principally for pre-heating prior to occupancy.

However, the designer is often faced with a situation in which there is little or no opportunity to influence important characteristics of the building that have a strong bearing on the heating system, particularly in the replacement of an existing heating system. For example, there may be some strains on the area and location of plant rooms, the space for and the routing of distribution networks. There may also be a requirement to interface with parts of an existing system, either for heating or ventilation. Where **domestic hot water**② is required, a decision is required on whether it should be heated by the same system as the space heating or heated at the point of use.

## 6. *Occupancy*

When the building is to be occupied, what activities are to be carried out within it are key determinants of the heating system specification. Are the occupants sedentary or physically active? What heat gains are expected to arise from processes and occupancy, including associated equipment such as computers and office machinery? Do all areas of the building have similar requirements or is there area with special requirements? These factors may determine or at least constrain the options available. The anticipated occupancy patterns may also influence the heating design at a later stage. Consideration should also be given to flexibility and adaptability of systems, taking account of possible re-allocation of floor space in the future.

## 7. *Energy efficiency*

The term "energy efficiency" gained currency during the 1980s and is now widely used.

In general the energy efficiency of a building can only be assessed in relative terms, either based on the previous performance of the same building or by comparison with other buildings. Thus the energy use of a building might be expressed in terms of annual energy use per square meter of floor area, and compared with benchmark levels for similar buildings. The result so obtained would depend on many physical factors including insulation, boiler efficiency, temperature, control systems, and the luminous efficacy of the lighting installations, but it would also depend on the way the occupants interacted with the building, particularly if it were naturally ventilated with openable windows.

The energy consumption of buildings is most readily measured in terms of "delivered" energy, which may be read directly from meters or from records of fuels bought in bulk. Delivered energy fails to distinguish between electricity and fuel which has yet to be converted to heat. "Primary" energy includes the overheads associated with production of fuels and with the generation and distribution of electricity. Comparisons of energy efficiency are therefore sometimes made on the basis of primary energy or on the emissions of "greenhouse" gases, which also takes account of energy overheads. Fuel cost may also be used and has the advantage of being both more transparent and more relevant to non-

---

② 见 Part 3 Extension.

technical building owners and occupants. In any event, it is meaningless to quote energy use in delivered energy obtained by adding electricity use to fuel use. Consequently, if comparisons are to be made in terms of delivered energy, electricity and fuel use must be quoted separately.

Clearly, the performance of the heating system has a major influence on energy efficiency, particularly in an existing building with relatively poor insulation. The designer has the opportunity to influence it through adopting an appropriate design strategy and choice of fuel, by specifying components with good energy performance, and by devising a control system that can accurately match output with occupant needs.

## 8. Make the strategic decisions

Each case must be considered on its own merits and rigorous option appraisal based on economic and environmental considerations should be undertaken. However, the flow charts shown in Fig. 5.3 and Fig. 5.4 are offered as general guidance. Fig. 5.3 refers to heating systems in general and Fig. 5.4 refers to choice of fuel.

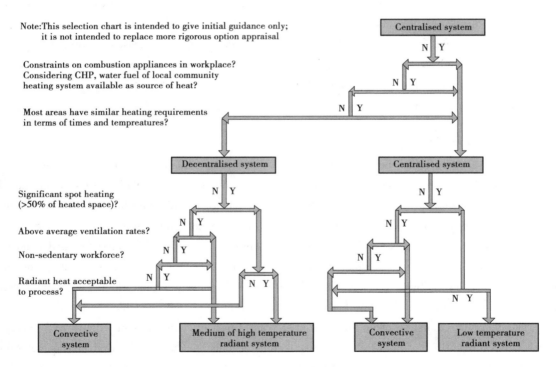

Fig. 5.3　Selection chart: heating systems

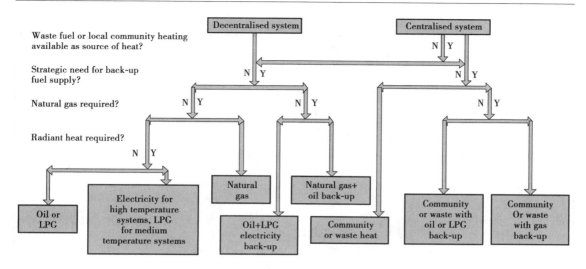

Fig. 5.4　Selection chart: Fuel

**Keywords**: Heating system, Building services, Energy efficiency, Integrated design

**Source from**:

CIBSE. Guide B Heating, ventilation, air conditioning and refrigeration. Section 1 Heating. London: CIBSE Publications, 2005.

# Part 3　Extension

1. Dry resultant temperature(干球综合温度)

干球综合温度[也称为"操作温度"(Operative temperature)]反映了环境空气温度和平均辐射温度的综合作用,可用于衡量人体热舒适,是根据人体与环境的干热交换(即以辐射和对流方式进行的热交换)推导出的温度参数。它假设人处于一个温度均匀的外壳中,人在其中通过辐射和对流方式进行与在实际环境中等量的干热交换。

2. Domestic hot water(生活热水)

生活热水,亦可称为民用热水或家用热水。合格的生活热水应该具备四大要素:水温、水压、水质、水量。生活热水的供应主要分为两类:集中供应生活热水类(例如太阳能集中生活热水系统);末端加热类(例如家用电热水器)。

# Part 4　Words and Expressions

| valve | *n.* 阀;真空管;(管乐器的)活栓 |
| --- | --- |
|  | *vt.* 装阀于;以活门调节 |
| building services | 建筑设备 |
| fabric | *n.* 织物;布;构造;(建筑物的)结构(如墙、地面、屋顶) |
| integrate | *v.* 使一体化;使整合;使完整 |

续表

| 英文 | 中文 |
| --- | --- |
| indoor temperature | 室内温度 |
| fuel costs | 燃料成本(消耗) |
| dry resultant temperature | 干球综合温度 |
| mean radiant temperature | 平均辐射温度 |
| thermal comfort | 热舒适 |
| condensation | n. 冷凝；冷凝液；凝结的水珠 |
| heating mains | 供热干管 |
| heat loss | 热损失 |
| orientation | n. 方向，定位，取向 |
| passive solar | 被动式太阳能 |
| legislation | n. 立法，制定法律；法律，法规 |
| obligation | n. 义务，责任；证券，契约；债务 |
| carbon performance rating | 碳性能评定(等级) |
| domestic hot water | 生活热水 |
| anticipate | vt. 预先考虑到；先于……前行动；提前支用 |
| energy efficiency | 能源效率 |
| luminous | adj. 发光的；明亮的；清楚的 |
| rigorous | adj. 严密的；缜密的；严格的；枯燥的 |
| thermal insulation | 绝热、保温 |
| infiltration | n. 渗透；下渗；渗滤；入渗 |
| ancillary | adj. 辅助的；补充的；附属的；附加的<br>n. 助手，随从 |
| damper | n. (钢琴的)制音器；减震器<br>adj. 潮湿的 |
| sensor | n. 传感器，灵敏元件 |
| ventilation | n. 空气流通；通风设备；通风方法 |
| heat recovery | 热回收 |

# Part 5  Reading

## Energy Performance in Heating System

The practical realization of energy efficiency depends not only on the characteristics of the equipment installed but also on how it is controlled and integrated with other equipment. The following parts describe aspects of energy efficiency that need to be taken into account in heating system design.

## 1. Thermal insulation

For new buildings, satisfying the Building Regulations will ensure that the external fabric has a reasonable and cost-effective degree of insulation (but not necessarily the economic optimum), and that insulation is applied to hot water storage vessels and heating pipes that pass outside heated spaces.

In existing buildings, considerations should be given to improving the thermal resistance of the fabric, which can reduce the heat loss significantly. This can offer a number of advantages, including reduced load on the heating system, improved comfort and the elimination of condensation on the inner surfaces of external walls and ceilings. In general, decisions on whether or not to improve insulation should be made following an appraisal of the costs and benefits, taking account both of running costs and the impact on capital costs of the heating system.

## 2. Reducing air infiltration

Infiltration can contribute substantially to the heating load of the building and cause discomfort through the presence of draughts and cold areas. As for fabric insulation, the costs and benefits of measures to reduce infiltration should be appraised on a life-cycle basis, taking account of both running costs and capital costs.

## 3. Seasonal boiler efficiency

Boiler efficiency is the principal determination of system efficiency in many heating system. What matters is the average efficiency of the boiler under varying conditions throughout the year, known as "seasonal efficiency". This may differ significantly from the bench test boiler efficiency, although the latter may be a useful basis for comparison between boilers.

Many boilers have a lower efficiency when operating at part load, particularly in an on/off control mode. Apart from the pre-heat period, a boiler spends most of its operating life at part load. This has led to the increased popularity of multiple boiler system since, at 25% of design load, it is better to have 25% of a number of small boilers operating at full output, rather than one large boiler operating at 25% output.

## 4. Efficiency of ancillary devices

Heating system rely on a range of electrically powered equipment to make them function, including pumps, fans, dampers, electrically actuated valves, sensors and controllers. Of these, pumps and fans are likely to consume the most energy, but even low electrical consumption may be significant if it is by equipment that is on continuously. It is important to remember that the cost per kWh of electricity is typically four times that of fuels used for heating, so it is important to avoid unnecessary electrical consumption.

For pumps and fans, what matters is the overall efficiency of the combined unit including the motor and the drive coupling. Fan and pump characteristics obtained from manufacturers should be used to design the system to operate around the point of maximum efficiency, taking account of both the efficiency of the motors and of the coupling to the pump or fan. Also, it is important that the drive ratios are selected to give a good match between the motor and the load characteristics of the equipment it is driving.

Pumping and fan energy consumption costs can be considerable and may be a significant proportion of total running costs in some heating systems. However, it may be possible to reduce running costs by specifying larger pipes or ductwork. Control system design can also have a significant impact on running costs. Pumps and fans should not be left running longer than necessary and multiple speed or variable speed drives should be considered where a wide flow range is required.

## 5. *Controls*

Heating system controls perform two distinct functions:

- They maintain the temperature conditions required within the building when it is occupied, including pre-heating to ensure that those conditions are met at the start of occupancy periods.
- They ensure that the system itself operates safely and efficiently under all conditions.

The accuracy with which the specified temperatures are maintained and the length of the heating period have a significant impact on energy efficiency and running costs. A poorly controlled system will lead to complaints when temperatures are low. The response may be raised set-points or extended pre-heat periods, both of which have the effect of increasing average temperatures and energy consumption. Controls which schedule system operation, such as boiler sequencing, can be equally important in their effect on energy efficiency, especially as the system may appear to function satisfactorily while operating at low efficiency.

## 6. *Zoning*

Rooms or areas within building may require to be heated to different temperatures or at different times, each requiring independent control. Where several rooms or areas of a building behave in a similar manner, they can be grouped together as a "zone" and put on the same circuit and controller. For instance, all similar south-facing rooms of a building may experience identical solar gain changes and some parts of the building may have the same occupancy patterns. The thermal responses of different parts of a zones, so that all parts of the zone reach their design internal temperature together. A poor choice of zones can lead to some rooms being too hot and others too cool.

## 7. *Ventilation heat recovery*

A mechanical ventilation system increases overall power requirements but offers potential energy savings through better control of ventilation and the possibility of heat recovery. The most obvious saving is through limiting the operation of the system to times when it is required, which is usually only when the building is occupied. The extent to which saving are possible depends crucially on the air leakage performance of the building. In a leaky building, heat losses through infiltration may be comparable with those arising from ventilation. In an airtight building, the heat losses during the pre-heat period may be considerably reduced by leaving the ventilation off and adopting a smaller plant size ratio.

Ventilation heat recovery extracts heat from the exhaust air for reuse within a building. It includes:

- "air-to-air" heat recovery, in which heat is extracted from the exhaust air and transferred to the supply air using a heat exchanger or thermal wheel;
- a heat pump, to extract heat from the exhaust air and transfer it to domestic hot water.

Air-to-air heat recovery is only possible where both supply air and exhaust air are ducted. High heat transfer efficiencies (up to 90%) can be achieved. Plate heat exchangers are favoured for use in houses and small commercial systems, while thermal wheels are typically used in large commercial buildings. Heat pipe systems offer very high heat efficiency and low running cost. Run-around coils may also be used and have the advantage that supply and exhaust air streams need not be adjacent to each other.

The benefits of the energy saved by heat recovery must take account of any additional electricity costs associated with the heat recovery system, including the effect of the additional pressure drop across the heat exchanger. Assessment of the benefits of heat recovery should also take account of the effect of infiltration, which may by-pass the ventilation system to a large extent. The cost-effectiveness of heat recovery also depends on climate and is greatest when winters are severe.

Heat pumps transferring heating from exhaust ventilation air to heat domestic hot water have widely been used in apartment buildings. The same principle has been successfully used in swimming pools.

**Source from:**
CIBSE. Guide B Heating, ventilation, air conditioning and refrigeration. Section 1 Heating. London: CIBSE Publications, 2005.

## Part 6　Exercises and Practices

## 1. *Discussion*

What does heat gain expected to arise from processes and occupancy, including associated equipment such as computers and office machinery?

## 2. *Translation*

(1) As the human body exchanges heat with its surroundings both by convection and by radiation, comfort depends on the temperature of both the air and the exposed surfaces surrounding it and on air movement.

(2) Aspects of the design and performance of heating systems are covered by building regulations aimed at the conservation of fuel and power and ventilation; and regulations set minimum efficiency levels for boilers.

(3) The result so obtained would depend on many physical factors including insulation, boiler efficiency, temperature, control systems, and the luminous efficacy of the lighting installations, but it would also depend on the way the occupants interacted with the building, particularly if it were naturally ventilated with openable windows.

(4) Infiltration can contribute substantially to the heating load of the building and cause discomfort through the presence of draughts and cold areas.

(5) Heating systems rely on a range of electrically powered equipment to make them function, including pump, fans, dampers, electrically actuated values, sensors and controllers.

## 3. Writing

Please explain how to calculate the heat load.

# Lesson 6

# Heating System Design

## Part 1  Preliminary

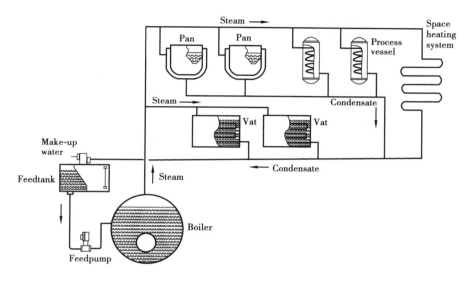

**Fig. 6.1  Typical steam heating circuit**

In a steam heating system, the heat is mainly derived from the latent heat of vaporization released by steam condensation. Since the latent heat of vaporization of steam is much greater than the heat released by the same mass of hot water, the steam mass flow is much smaller than the hot water flow with the same heat load. According to the different backwater hydrodynamic force, the steam heating system can be divided into gravity, residual pressure and mechanical circulation. Figure 6.1 shows the mechanical steam heating circulation forced by a feed-pump, which consists of condensate tanks, air pipes, steam traps, boilers and heat emitters.

## Part 2  Text

热水供暖系统由热源、输配系统和散热设备组成。热水供暖系统有多种类型:按热媒温度,可分为高温水系统(水温大于100 ℃)和低温水系统(水温小于等于100 ℃),在实际运行中合理降低供暖系统的热媒温度,有利于提高散热器供暖的热舒适和节能降耗;按介质循环动力,可分为重力循环系统和机械循环系统,与前者相比,后者的作用范围更大,应用更广泛;按热水介质是否接触大气,分为开式系统和闭式系统;按管线结构形式可分为垂直式和水平式,单管和双管式,上分式、下分式和中分式,同程式和异程式。

Hydronic systems use hot water for transferring heat from the heat generator to the heat emitters. The most usual type of heat generator for hydronic systems is a "boiler", misleadingly named as it must be designed to avoid boiling during operation. Hot water may also be generated by **heat pumps**①, and waste heat can be reclaimed from processes and by solar panels, the latter of which typically being used to produce domestic hot water in summer. Heat emitters take a variety of forms including panel radiators, natural and forced convectors, fan-coil units, and under-floor heating. Hydronic systems normally rely on pumps for circulation, although gravity circulation was favored for systems design before around 1950.

Hydronic systems offer considerable flexibility in type and location of emitters. The heat output available in radiant form is limited by the temperature of the circulation water. But for radiators and heated panels, can be sufficient to counteract the effect of cold radiation from badly insulated external surfaces. Convective output can be provided by enclosed units relying on either natural or forced air-convection. Flexibility of location is ensured by the small diameter of the circulation pipework and the wide variety of emitter sizes and types.

In addition to the sizing of emitters and boilers, the design of hydronic systems involves the hydraulic design of the circulation system to ensure that water reaches each emitter at the necessary flow rate and that the pressures around the system are maintained at appropriate levels. System static pressures may be controlled either by sealed expansion vessels or by hydrostatic pressure arising from the positioning of cisterns at atmospheric pressure above the highest point of the circulating system. Both cisterns and pressure vessels must cope with the water expansion that occurs as the system heats up from cold; the design of feed, expansion and venting is crucial to both the safety and correct operation of systems.

## 1. *Operating temperatures for hydronic systems*

The operating temperature of a hydronic heating system both determines its potential performance and affects its design. Systems are generally classified according to the temperature and static pressure at which they operate. Low pressure hot water (LPHW) systems may be either sealed or open to the atmosphere and use a variety of materials for the distribution pipework. Also, the operating temperature should be set low enough that exposed heat emitters, such as panel radiators, to prevent a burn hazard to building occupants. Medium and high pressure systems are favored where a high heat output is required, such as in a fan coil system in a large building. High pressure systems are particularly favored for distribution mains, from which secondary systems extract heat by heat exchangers for local circulation at lower temperatures.

The efficiency of a condensing boiler is more strongly influenced by the return temperature, rather than the flow temperature, which ought to be a further encouragement to use large values of $(t_1 - t_2)$. However, a larger temperature difference lowers the mean water temperature of the emitter, which reduces specific output and requires larger surface area.

---

① 见 Part 3 Extension.

In general, it may be noted that output tends to increase disproportionately as the difference between the mean system temperature and the room temperature increase. This favors the use of a high system temperature. However, other factors need to be considered which may favor a lower temperature, including the surface temperature of radiators, boiler operating efficiency and the characteristics of certain heat emitters. For example, under-floor heating is designed to operate with low system temperatures to keep floor surface temperatures below 29 ℃.

## 2. *System layout and design*

Systems must be designed to match their specified design heat load, including domestic hot water provision where required, and to have controls capable of matching output to the full range of variation in load over a heating season. Separate circuits may be required to serve zones of the building with different heat requirements. In addition, there must be provision for hydraulic balancing of circuits and sub-circuits, and for filling, draining and venting of each part of the system.

Distribution systems may be broadly grouped into one-pipe and two-pipe categories. In one-pipe systems, radiators are effectively fed in series, and system temperature varies around the circuit. Control of one-pipe systems requires setting of by-passes and 3-port valves. Two-pipe systems operate at nominally the same temperature throughout the circuit but require good balancing for that condition to be achieved in practice. Control of two-pipe systems may employ either 2-port or 3-port valves to restrict flow to individual heat emitters.

### Hydraulic design

Hydraulic design needs to take account of the effect of water velocity on noise and erosion, and of the pressure and flow characteristics of the circulation pump.

Pumps should be capable of delivering the maximum flow required by the circuit at the design pressure drop around the circuit of greatest resistance, commonly known as the index circuit. If variable speed pumping is to be used, the method of controlling pump speed should be clearly described and the pump should be sized to operate around an appropriate part of its operating range. The location and sizing of control valves need to take account of pressure drops and flows around the circuit to ensure that they operate with sufficient valve authority.

### Balancing

The objective of balancing is to ensure that each emitter receives the flow required at the design temperature. Balancing may be carried out most precisely by measuring and adjusting flow to individual parts of the circuit, but can also be carried out by observing temperatures throughout the system. Temperature-based balancing is commonly used on domestic systems but has the disadvantage that the adjustments must be made and checked when the system has reached a steady-state, which may take a considerable time.

It is important to take account of the need for balancing at the design stage, including the location of measuring stations around the system, the equipment needed to achieve balancing, and the procedures for carrying it out. Balancing by flow requires a provision for flow measurement and, in all cases, appropriate valves must be installed to control the flow to particular parts of the circuit. Balancing procedures, including a technical specification for commissioning the system, and the responsibilities of the various parties involved should be clearly identified at the outset.

The design of pipework systems can have a considerable effect on the ease with which balancing can be achieved. Reverse return circuits, which ensure that each load has a similar circuit length for its combined flow and return path, can eliminate much of the inequality of flow that might otherwise need to be rectified during balancing. Distribution manifolds and carefully selected pipe sizes can also assist with circuit balancing. It is important to avoid connecting loads with widely differing pressure drops and heat emitting characteristics (e.g. panel radiators and fan coil units) to the same sub-circuit.

## 3. *Choice of heat source*

The choice of heat source will depend on the options available. These are outlined below.

### Boilers

Boilers are available in a large of types and sizes and, unless they are connected to a community heating system and almost all hydronic heating systems rely on one or more boilers. Boiler efficiency has improved markedly over the past two decades. Technical developments have included the use of new materials to reduce water content and exploit the condensing principle, gas-air modulation to improve combustion efficiency and modularization to optimize system sizing. These developments have resulted in considerable improvements in performance at part load, with considerable benefit to seasonal efficiency.

**Condensing boilers**[②] have efficiencies of up to 92% (gross calorific value) and are no longer much more expensive than other boilers. Neither are they so widely differentiated from non-condensing boilers in their performance, as the latter have improved considerably in their efficiency. Seasonal efficiency is the principal characteristic affecting the running cost of a boiler (or boiler system). In considering whole life cost, the lifetime of components should be taken into account.

"Combination" boilers provide an instantaneous supply of domestic hot water in addition to the usual boiler function. Their main advantage lies in the space they save, as they need no hot water storage cylinder or associated storage cistern. Also, they typically incorporate an expansion vessel for sealed operation, so that they need no plumbing in the loft space; this is particularly advantageous in flats where it may be difficult to obtain sufficient head from an open system. A further advantage is the elimination of heat losses from the hot water stored in the cylinder. Combination boilers have gained a large share of the market for boilers installed in housing over the past decade. However, the limitations of combination boilers should also be understood by both the installer and the client. The maximum flow rate at which hot water can be draw is limited, especially over a prolonged period

---

② 见 Part 3 Extension.

or when more than one point is being served simultaneously. Combination boilers are also susceptible to scaling by hard water, as the instantaneous water heating function requires the continual passage of water direct from the mains through a heat exchanger.

## Heat pumps

Heat pumps have a number of different forms and exploit different sources of low grade heat. Worldwide, the heat pumps most widely used for heating are reversible air-to-air units that can also be used for cooling. Such units are typically found where there is significant need for cooling and the need for heating is limited. In the China climate, electrically driven air-to-air heat pumps are not frequently installed solely to provide heating, which may be explained by the relatively high price of electricity in relation to gas. Heat pumps offer a particularly attractive option for heating when there is a suitably large source of low grade heat, such as a river, canal or an area of ground. Gas-fired ground source heat pumps currently being evaluated for use in housing as a boiler replacement are reported to have a seasonal coefficient of performance of around 1.4.

## Solar panels

Solar water heating panels are widely used around the world to provide domestic hot water, particularly where sunshine is plentiful and fuel is relatively expensive, but are rarely used for space heating. In the typical climate, a domestic installation can typically provide hot water requirements for up to half the annual hot water requirements, using either a separate pre-heat storage cylinder or a cylinder with two primary coils, one linked to the solar panel and the other to a boiler. Although technically successful, the economics of such systems are at best marginal when assessed against heat produced by a gas or oil boiler and they are rarely used in nondomestic buildings. Solar panels are also widely used for heating outdoor swimming pools in summer, for which they are more likely to be cost effective.

## Community heating

If available, consideration should be given to utilizing an existing supply of heat from a district or local heat supply ("community heating"). Heat supplied in this way may be of lower cost and may also have significantly lower environmental impact, especially if it is generated using **combined heat and power (CHP)**[3] or makes use of heat from industrial processes or waste combustion. The low net $CO_2$ emissions from heat from such sources can contribute significantly to achieving an environmental target for a building.

## Stand-alone CHP systems

Where there is no suitable existing supply of heat, the opportunity for using a stand-alone combined heat and power (CHP) unit should be evaluated. The case for using CHP depends on require-

---

③ 见 Part 3 Extension.

ments both for heat and electricity, their diurnal and seasonal variability and the extent to which they occur simultaneously. The optimum CHP plant capacity for a single building needs to be determined by an economic assessment of a range of plant sizes and in general will result in only part of the load being met by CHP, the rest being met by a boiler. It is important to have a reasonable match between the generated output and electricity demand, as the value of the electricity generated tends to dominate the economic analysis; the optimum ratio of heat demand to power demand generally lies between 1.3:1 and 2:1. There may be opportunities for exporting electricity. The best price for exported electricity is likely to be obtained from consumers who can link directly to the system rather than from a public electricity supplier. Where standby power generation is required to reduce dependency of public supplies of electricity, it may be particularly advantageous to install a CHP unit, thereby avoiding the additional capital cost of a separate standby generator.

## 4. Choice of heat emitter

Hydronic systems are capable of working with a wide variety of heat emitters, offering a high degree of flexibility in location, appearance and output characteristics. This section deals with some of the principal characteristics of emitters affecting their suitability for particular situations.

### Radiators

Radiators, usually of pressed steel panel construction, are the most frequent choice of emitter. They are available in a wide variety of shapes, sizes and output ranges, making it possible to obtain a unit (or units) to match the heat requirements of almost any room or zone.

Despite their name, radiators for hydronic system usually produce more than half their surface area.

### Natural convectors

Wall-mounted natural convectors may be used instead of radiators. They may be used where is insufficient space for mounting radiators, for example in base-board or trench heating configurations. The output from natural convectors varies considerably with design and manufacturer's data for individual emitter types should be used.

### Fan coil heaters

Fan coil units produce high heat outputs from compact units using forced air circulation. Their output may be considered to be entirely convective and is approximately proportional to temperature difference. Where systems contain a mixture of natural and forced air appliances, the different output characteristic of the two types should be taken into account, particularly with regard to zoning for control systems.

### Floor heating

Floor heating (also referred to as under-floor heating) uses the floor surface itself as a heat emitter. Heat may be supplied either by embedded electric heating elements or by the circulation of water as a part of hydronic system, involving appropriately spaced pipes positioned beneath the floor surface. The pipes may be embedded within the screed of a solid floor or laid in a carefully controlled configuration beneath a suspended floor surface. Insulation beneath the heating elements is clearly very important for good control of output and to avoid unnecessary heat loss.

The heat emission characteristics of floor heating differ considerably from those radiator heating. Floor surface temperature is critical to comfort, as well as to heat output. The optimum floor temperature range for control lies between 21 and 28 ℃ depending on surface material, so systems are normally designed to operate at no higher than 29 ℃ in occupied areas. Higher temperatures are acceptable in bathrooms and close to external walls with high heat loss, such as beneath full-length windows.

The design surface temperature is controlled by the spacing between pipes and the flow water temperature. It is also affected by floor construction, floor covering and the depth of the pipes beneath the floor surface; detailed design procedures are given by system manufacturers. In practice, systems are usually designed to operate at flow temperatures of between 40 and 50 ℃, with a temperature drop of between 5 and 10 ℃ across the system. Maximum heat output is limited by the maximum acceptable surface temperature to around 100 $W/m^2$ for occupied areas.

Floor heating may be used in conjunction with radiators, for example, for the ground floor of a house with radiators on upper floors. Separate circuits are required in such cases, typically using a mixing valve to control the temperature of the under-floor circuit. Floor heating is best suited to well insulated buildings, in which it can provide all the required heating load.

## 5. *Pumping and pipework*

The hydraulic requirements for a system are derived from parameters such as system operating temperature and the heat output required from emitters, which affect pipework layout. The design also needs to take account of the effect of water velocity on noise and corrosion, and the pressure and flow characteristics required of the circulation pump. The key design decisions include:
- System pressures;
- Whether to use an open or a sealed pressurization method;
- Which material to use for pipes;
- The flow velocity to be used;
- How the system is to be controlled;
- Filling and air removal arrangements;
- Pumping requirements, i.e. variable or fixed flow rate.

## 6. *Energy storage*

Energy storage may either be used to reduce peak loads or to take advantage of lower energy prices at certain times of day. Heat is stored using either solid cores or hot water vessels. The most common application of thermal storage is in dwellings, in which solid core storage is charged with heat at off-peak rates for a 7 or 8 hour period.

Systems relying on hot water storage vessels are also available for use in dwellings. The three main types are as follows:

——Combined primary storage units (CPSU): provide both space and water heating from within a single appliance, in which a burner heats a thermal store. The water in the thermal store is circulated to radiators to provide space heating, while a heat exchanger is used to transfer heat to incoming cold water at mains pressure to provide a supply of domestic hot water.

——Integrated thermal stores: also provide both space and water heating from within a single appliance. However, they differ from CPSUs in that a separate boiler is used to heat the primary water.

——Hot-water-only thermal stores: use thermal storage only for production of domestic hot water. As for the two types described above, the domestic hot water is provided by a heat exchanger working at mains pressure.

Thermal storage for larger buildings must rely on purpose-designed storage vessels with capacity and storage temperature optimised for the heat load. Other design parameters that must be considered are insulation of the storage vessel, arrangements for dealing with expansion and the control strategy for coupling the store to the rest of the system.

## 7. *Domestic hot water*

Whether or not to produce domestic hot water from the same system as space heating is a key decision to be taken before detailed design proceeds. In housing, where demand for hot water is a substantial proportion of the total heat load, a hydronic heating system is usually the most convenient and satisfactory means of producing hot water, using either a hot water storage cylinder or a combination boiler.

In buildings other than housing, the case for deriving domestic hot water from a hydronic heating system depends greatly on circumstances. The demand for hot water and the locations with the building where it is required will affect the relative costs of independent heat generation and connection to the space heating system. In general, independent hot water generation is the most economical choice when relatively small amounts of hot water is required at positions distant from the boiler. Circulating hot water circuits that require long pipe runs and operate for extended periods solely to provide hot water can waste large amounts of energy, particularly during summer months when no space heating is required. In commercial buildings, toilet areas are often best served by independent gas or electric water heaters.

## 8. *Control for hydronic systems*

Hydronic heating systems are capable of very close control over environmental conditions using a range of strategies. The choice of control system type will depend on the closeness of control required, the number of different zones that must be controlled independently and the times at which the building will be occupied and require heating. The design must also take account of the characteristics of both heat generators and emitters.

A typical control system for a hydronic heating system in a dwelling or small building consists of a programmer, which may incorporate a time-switch or optimum start/stop functions, a room thermostat for each zone, motorised valves to control the flow to each zone and, if necessary, a frost protection thermostat. Where domestic hot water is also provided by the system, a thermostat and motorised valve to control the temperature of the hot water storage cylinder are also needed. Controls should be wired in such a way that the boiler operates only when a space heating or cylinder thermostat is calling for heat. Thermostatic radiator valves (TRV) may be used to control individual rooms within a zone. Pump "over-run" (i.e. delay in switching off a pump) may also be provided by the system or may be incorporated in the boiler controls.

Hydronic systems in larger buildings are likely to have more complex controls, including optimum start, and often incorporate weather compensation in which the system flow temperature is controlled in response to external temperature, according to a schedule derived for the building. Where there are multiple or modular boilers, sequence control is required for the boilers. Variable speed pumping may also be used. The pump speed is usually controlled to maintain a constant pressure differential across a point in the circuit as flow reduces in response to 2-port valve and TRV positions. Care is needed in the choice of valves used for control to ensure good "valve authority", which means that they are sized appropriately in relation to the pressure drops around the circuit.

**Keywords**: Hydronic system, Heat generator, Hydaulic design, Heat emitter

**Source from**:
CIBSE. Guide B Heating, ventilation, air conditioning and refrigeration. Section 1 Heating. London: CIBSE Publications, 2005.

# Part 3　Extension

1. Heat pumps（热泵）

热泵技术是近年来在全世界备受关注的新能源技术。人们所熟悉的"泵"是一种可以提高位能的机械设备，比如水泵主要是将水从低位抽到高位；而"热泵"是一种能从自然界的空气、水或土壤中获取低品位热能，经过电力做功，提供可被人们所用的高品位热能的装置。

按热源种类可分为：水源热泵、地源热泵、空气源热泵、双源热泵（水源热泵和空气源热泵结合）。

2. Condensing boilers（冷凝锅炉）

冷凝锅炉就是利用高效的烟气冷凝余热回收装置来吸收锅炉尾部排烟中的显热和水蒸气凝结所释放的潜热，以达到提高锅炉热效率的目的。

燃料燃烧会产生大量的 $CO_2$、$NO_x$ 和少量的 $SO_2$，这些物质排放到大气中会引起温室效应和产生酸雨，对环境产生破坏作用。冷凝锅炉在冷凝烟气中水蒸气的同时，可以方便地去除烟气中的有害物质，因此，采用冷凝锅炉对保护环境也具有重要的意义。

3. Combined heat and power（热电联产）

Combined heat and power，简称 CHP。发电厂既生产电能，又利用汽轮发电机做过功的蒸汽对用户供热的生产方式，是指同时生产电、热能的工艺过程，较之分别生产电、热能方式节约燃料。以热电联产方式运行的火电厂称为热电厂。对外供热的蒸汽，压力通常分为 0.78~1.28 MPa 和 0.12~0.25 MPa 两种。前者供工业生产，后者供民用采暖。热电联产的蒸汽没有冷源损失，所以能将热效率提高到 85%，比大型凝汽式机组（热效率达 40%）还要高得多。

## Part 4　Words and Expressions

| | | |
|---|---|---|
| make-up water | | 补给水 |
| steam heating circuit | | 蒸汽供暖环路 |
| hydronic system | | 水系统 |
| heat emitters | | 散热器 |
| boiler | | 锅炉 |
| heat pump | | 热泵 |
| panel radiators | | 板式散热器 |
| fan-coil units | | 风机盘管 |
| under-floor heating | | 地板采暖 |
| gravity circulation | | 重力循环 |
| counteract | *vt.* | 抵消；阻碍；中和 |
| forced air-convection | | 强制对流 |
| cistern | *n.* | 蓄水池，储水箱 |
| pressure vessel | | 压力罐 |
| static pressure | | 静压 |
| extract | *vt.* | 提取；（费力地）拔出；选取 |
| condensing boiler | | 冷凝式锅炉 |
| temperature difference | | 温度差 |
| disproportionately | *adv.* | 不匀称，不相称 |
| provision | *n.* | 规定，条项，条款；预备，设备；供应 |
| | *vt. & vi.* | 为……提供所需物品 |
| hydraulic design | | 水力计算 |
| erosion | *n.* | 腐蚀，侵蚀，磨损 |
| resistance | *n.* | 抵抗；阻力；抗力；电阻 |
| variable speed pump | | 变频泵 |
| reverse | *adj.* | 反面的；颠倒的；倒开的 |
| | *n.* | 倒转，反向；倒退；失败 |
| manifold | *adj.* | 多种多样的 |
| | *n.* | 具有多种形式的东西；多支管；歧管 |
| modulation | *n.* | 调制；调幅度；移调 |
| combustion | *n.* | 燃烧，烧毁；氧化；骚动 |
| instantaneous | *adj.* | 瞬间的；即刻的；猝发的 |
| susceptible | *adj.* | 易受影响的；易受感染的 |
| combined heat and power (CHP) | | 热电联产 |
| wall-mounted | | 壁挂式的 |

## Part 5    Reading

## District Heating

### 1. *What is district heating*?

District heating is an infrastructure which allows heat generated in a centralized location to be distributed to residential homes and commercial buildings in a larger area. District heating is clean, efficient and cost-effective due to its flexibility and optimal heat generation conditions.

A district heating plant is often a combined heat and power (CHP) plant. By co-producing heat and power in the same process, the heat that would otherwise be wasted in electricity production is utilized. This leaves tremendous energy savings of up to 30%.

In essence, any energy source can be used in a district heating system. Renewables such as biomass, solar energy and waste are becoming increasingly applied in district heating utilities either completely or as a complement to traditional fossil fuels. District heating is in other words ready for today's clean energy sources as well as those of the future.

### 2. *How does district heating works*?

Through a district heating network, the heat producing plant pumps heated supply water to consumers where it is used as room-/floor-heating and to generate domestic hot water. The domestic hot water gets heated in a heat exchanger in which the heated supply water transfers its heat to the water coming out of the taps.

For room heating, the supply water might be used directly. Alternatively, a heat exchanger could also transfer the heat to an internal circulation. The supply water, which is now cold because the heat has been transferred to domestic hot water and room heating, then returns to the district heating plant. The district heating supply water circulates endlessly in a closed pipeline.

Some district heating systems use steam as medium for heat distribution instead of water. This is to achieve higher supply temperatures, which are often necessary for industrial processes. A disadvantage of steam is that it has higher heat losses than water.

### 3. *How a heat supply network operates*?

In a heat supply network, thermal energy can be transported many kilometers by a producer to a consumer. Transport of this heat energy ensues via strongly insulated piping systems.

Purified water is generally used as a formation fluid, due to its roughly specific heat capacity. Hot water is transported to the consumer via the flow line. There, heat is extracted from the water through a heat exchanger and the water stays in the pipes. The cooled water then flows back to the heat generator via the return line, where it is once again heated. The heat supply network therefore constitutes a closed cycle system.

Pre-insulated pipes are exclusively used for such a piping system. This means that the insulation has already been provided by the manufacturer of the pipes and they only need to be installed on site. It has been shown that this has resulted in a reduction of manufacturing and installation costs as well as an improvement in the level of insulation and the effectiveness of the heat supply network.

Pipes with an outer casing of polyethylene (PE) and insulation of polyurethane rigid foam (PUR) have proven to be the best technical and economical choice in this case. Combined with a long service life and long-term maintenance-free use of the pipes, in addition to cost-efficient installation on a bed of sand, the operation of expansive networks is made possible.

The exact configuration of the pipes varies according to the temperature of the water and the on-site requirements. Within a range of up to 140 ℃, rigid pre-insulated pipes are mostly used.

For low temperatures, flexible piping systems, which are delivered on reels, can be used as an alternative in order to save on investment costs. As an example, flexible pipes can be used as a medium pipe made from one of the following materials:

- soft steel;
- cross-linked polyethylene;
- copper.

## 4. Cogeneration

Cogeneration is the term given to the combined production of electricity and heat. In thermal power stations, only a part of the primary energy used can be converted into electricity. The excess electricity is designated as surplus or waste heat. In conventional power stations, this waste heat has to be completely dissipated into the surrounding environment.

Through cogeneration, a large part of this energy can be used as district heating, whereby the thermal output is emitted between the turbine stages. As a result, this enables the power station to use over 90% of its entire degree of efficiency. Heat supply networks bring heat to consumers, where it can be used to heat rooms and water.

As thermal electricity production can only be produced in an energy efficient way through expedient use of waste heat, cogeneration has subsequently become an essential part of modern energy supply in many countries.

## 5. Environmentally friendly & energy-efficient

Due to simultaneous production of heat and electricity in combined heat and power plants, district heating is very energy-efficient. By implementing renewable energy sources and utilizing waste heat generated by industry the environmental gain by district heating is further evident. This kind of energy utilization is beneficial for both environment and society in general.

Compared to individual heating systems the district heating plants are better at reducing emissions of hazardous compounds since they have more advanced pollution control equipment and through their more controlled conditions when generating heat. Furthermore, district heating is very convenient for consumers, who hardly notice how their radiators and tap water is heated in their everyday.

## 6. *Primary energy savings through thermal storage in district heating networks*

District heating is an efficient way to provide heat to residential, tertiary and industrial users. Heat is often produced by CHP (combined heat and power) plants, usually designed to provide the base thermal load (40% ~ 50% of the maximum load) while the rest is provided by boilers. The use of storage tanks would permit to increase the annual operating hours of CHP: heat can be produced when the request is low (for instance during the night), stored and then used when the request is high. The use of boilers results partially reduced and the thermal load diagram is flattered. Depending on the type of CHP plant this may also affect the electricity generation. All these considerations are crucial in the free electricity market.

## 7. *Steam, warm air & radiation systems*

### Steam systems

Steam systems use dry saturated steam to convey heat from the boiler to the point of use, where it is released by condensation. Control of heat output is generally by variation of the steam saturation pressure within the emitter. The resulting condensate is returned to the feed tank, where it becomes a valuable supply of hot feed-water for the boiler. The flow of steam is generated by the pressure drop that results from condensation. Condensate is returned to the lowest point in the circuit by gravity.

Steam offers great flexibility in application and is long established as a medium for heating in buildings. However, it is not frequently chosen as a medium for heating buildings when that is the sole requirement. It is much more likely to be appropriate when there are other requirements for steam, such as manufacturing processes or sterilisation. In such cases, steam may be the most satisfactory medium both for space heating and for domestic hot water generation. In many cases, it will be appropriate to use steam to generate hot water in a heat exchanger for distribution in a standard hydronic heating system.

### Warm air systems

Warm air heating can be provided either by stand-alone heaters or distributed from central air-handling plant; in many cases the same plant is used for summertime cooling/ventilation. Almost all the heat output is provided in convective form so the room air temperature is usually greater than the dry resultant temperature. Warm air systems generally have a much faster response time than hydronic systems. For example, a typical factory warm air system will bring the space up to design temperature within 30 minutes. Warm air systems can cause excessive temperature stratification, with warm air tending to collect at ceiling level. This may be particularly unwelcome in buildings with high ceilings, although it can be overcome by the use of de-stratification systems.

Warm air systems may be used to provide full heating to a space or simply supply tempered "make-up" air to balance the heat loss and air flow rate from exhaust ventilation systems. A slight excess air flow can be used to pressurize the heated space slightly and reduce cold draughts.

**Radiation systems**

In general, systems are considered to be radiant when more than 50% of their output is radiant, which corresponds broadly to those with emitter temperatures greater than 100 ℃. This definition includes medium temperature systems, such as high pressure hydronic systems, steam systems and air heated tubes, which operate at temperatures up to 200 ℃. High temperature radiant systems, such as those with electric radiant elements or gas heated plaques, produce a higher proportion of their output in radiant form and are particularly effective when heat output needs to be focused and directed to specific locations.

Radiant heating is particularly useful in buildings with high air change rates or large volumes that do not require uniform heating throughout, e.g., factories, and intermittently heated buildings with high ceilings. The key characteristics of radiant heating are as follows:

——Heat transfer occurs by radiation directly on surfaces, including building occupants and the internal surfaces of buildings and fittings. The surrounding air need not be heated to the same temperature as would be required with convective heating.

——A rapid response can be achieved because the effect of the thermal inertia of the building is by passed by direct radiation.

——After an initial warm-up period, radiant heating directed downwards towards floor level is augmented by re-radiation and convection from surfaces at the level occupied by people.

——Radiant asymmetry is a potential problem and may place restrictions on design.

Radiant heating can require less energy than convective heating because it enables comfort conditions to be achieved at lower air temperatures. As a general rule it is likely to have an advantage in this respect whenever ventilation heat losses exceed fabric heat losses. Further savings may be achieved when only some zones within a large open area require heating and local radiant temperature can be raised by well directed radiant heat. In such cases, large volumes of surrounding air may be left at much lower temperatures without a detrimental effect on dry resultant temperature in the working zones.

**Source from:**

[1] CIBSE. Guide B Heating, ventilation, air conditioning and refrigeration. Section 1 Heating. London: CIBSE Publications, 2005.

[2] Lund H, Möller B, Mathiesen BV, Dyrelund A. The role of district heating in future renewable energy systems. Energy 2010, 35:1381-1390.

# Part 6　Exercises and Practices

## 1. Discussion

(1) How does the CHP system operate?

(2) How do you choose the heat emitters?

## 2. Translation

(1) Hot water may also be generated by heat pumps, waste heat reclaimed from processes and by solar panels, the latter typically being used to produce domestic hot water in summer.

(2) High pressure systems are particularly favored for distribution mains, from which secondary systems extract heat exchangers for local circulation at lower temperatures.

(3) Solar water heating panels are widely used around the word to provide domestic hot water, particularly where sunshine is plentiful and fuel is relatively expensive, but are rarely used for space heating.

(4) Where systems contain a mixture of natural and forced air applications, the different output characteristic of two types should be taken into account, particularly with regard to zoning for control systems.

(5) A typical control system for a hydronic heating system in a dwelling or small building consists of a programmer, which may incorporate a time-switch or optimum start/stop functions, a room thermostat for each zone, motorised valves to control the flow to each zone and, if necessary, a frost protection thermostat.

## 3. Writing

Please explain the advantages & disadvantages of radiators, comparing with natural convectors or fan coil heaters.

# Lesson 7
# Air Conditioning Fundamentals

## Part 1  Preliminary

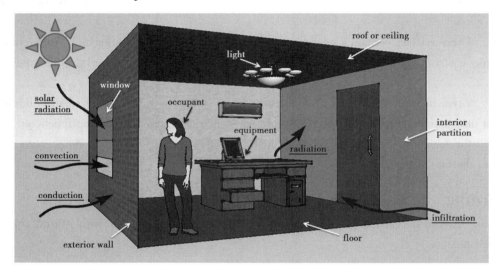

Fig. 7.1  Space heat gain

A cooling load calculation determines total sensible cooling load from heat gain (1) through opaque surfaces (walls, floors, ceilings, and doors), (2) through transparent fenestration surfaces (windows, skylights, and glazed doors), (3) caused by infiltration and ventilation, and (4) because of occupancy. The latent portion of the cooling load is evaluated separately. Although the entire structure may be considered a single zone, equipment selection and system design should be based on room-by-room calculations. For proper design of the distribution system, the conditioned airflow required by each room must be known.

## Part 2  Text

在空调负荷设计中,得热量、冷负荷、除热量和盘管换热量是不容混淆的几个概念。由于辐射散热转化为负荷的延迟效应,故得热量和冷负荷不一定相等。当空调系统连续运行并使室温维持相对恒定时,除热量就等于冷负荷。房间负荷计算是空调设计的基本环节。建筑构造及材料、房间实际运用方式、用户的行为习惯以及周边环境等因素,均会影响计算负荷的准确性。因此,负荷计算需要详细的建筑设计信息、室外设计条件、室内设计参数、室内热源特性及使用时间等数据。

Cooling loads result from many conduction, convection, and radiation heat transfer processes

through the building envelope and from internal sources and system components. Building components or contents that may affect cooling loads include the following:
- **External**: walls, roofs, windows, skylights, doors, partitions, ceilings, and floors;
- **Internal**: lights, people, appliances, and equipment;
- **Infiltration**: air leakage and moisture migration;
- **System**: outside air, duct leakage and heat gain, reheat, fan and pump energy, and energy recovery.

# 1. *Terminology*

The variables affecting cooling load calculations are numerous, often difficult to define precisely, and always intricately interrelated. Many cooling load components vary widely in magnitude, and possibly direction, during a 24 h period. Because these cyclic changes in load components often are not in phase with each other, each component must be analyzed to establish the maximum cooling load for a building or zone. A zoned system (i.e., one serving several independent areas, each with its own temperature control) needs to provide no greater total cooling load capacity than the largest hourly sum of simultaneous zone loads throughout a design day; however, it must handle the peak cooling load for each zone at its individual peak hour. At some times of day during heating or intermediate seasons, some zones may require heating while others require cooling. The zones' ventilation, humidification, or dehumidification needs must also be considered.

## Heat flow rates

In air-conditioning design, the following four related heat flow rates, each of which varies with time, must be differentiated.

**Space heat gain**. This instantaneous rate of heat gain is the rate at which heat enters into and/or is generated within a space. Heat gain is classified by its mode of entry into the space and whether it is sensible or latent. Entry modes include (1) solar radiation through transparent surfaces; (2) heat conduction through exterior walls and roofs; (3) heat conduction through ceilings, floors, and interior partitions; (4) heat generated in the space by occupants, lights, and appliances; (5) energy transfer through direct-with-space ventilation and infiltration of outdoor air; and (6) miscellaneous heat gains. Sensible heat is added directly to the conditioned space by conduction, convection, and/or radiation. Latent heat gain occurs when moisture is added to the space (e.g., from vapor emitted by occupants and equipment). To maintain a constant humidity ratio, water vapor must condense on the cooling apparatus and be removed at the same rate it is added to the space.

**Radiant heat gain**. Radiant energy must first be absorbed by surfaces that enclose the space (walls, floor, and ceiling) and objects in the space (furniture, etc.). When these surfaces and objects become warmer than the surrounding air, some of their heat transfers to the air by convection. The composite heat storage capacity of these surfaces and objects determines the rate at which their respective surface temperatures increase for a given radiant input, and thus governs the relationship between the radiant portion of heat gain and its corresponding part of the space cooling load.

The thermal storage effect is critical in differentiating between instantaneous heat gain for a given space and its cooling load at that moment.

**Space cooling load**. This is the rate at which sensible and latent heat must be removed from the space to maintain a constant space air temperature and humidity. The sum of all space instantaneous heat gains at any given time does not necessarily (or even frequently) equal the cooling load for the space at that same time.

**Space heat extraction rate**. The rates at which sensible and latent heat are removed from the conditioned space equal the space cooling load only if the room air temperature and humidity are constant. Along with the intermittent operation of cooling equipment, control systems usually allow a minor cyclic variation or swing in room temperature; humidity is often allowed to float, but it can be controlled. Therefore, proper simulation of the control system gives a more realistic value of energy removal over a fixed period than using values of the space cooling load. However, this is primarily important for estimating energy use over time; it is not needed to calculate design peak cooling load for equipment selection.

**Cooling coil load**. The rate at which energy is removed at a cooling coil serving one or more conditioned spaces equals the sum of instantaneous space cooling loads (or space heat extraction rate, if it is assumed that space temperature and humidity vary) for all spaces served by the coil, plus any system loads. System loads include fan heat gain, duct heat gain, and outdoor air heat and moisture brought into the cooling equipment to satisfy the ventilation air requirement.

## Time delay effect

Energy absorbed by walls, floor, furniture, etc., contributes to space cooling load only after a time lag. Some of this energy is still present and reradiating even after the heat sources have been switched off or removed, as shown in Figure. 7.2.

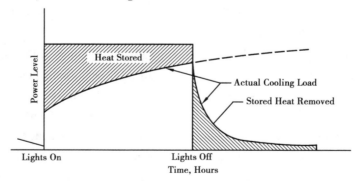

**Fig. 7.2 Thermal storage effect in cooling load from lights**

There is always significant delay between the time a heat source is activated, and the point when reradiated energy equals that being instantaneously stored. This time lag must be considered when calculating cooling load, because the load required for the space can be much lower than the instantaneous heat gain being generated, and the space's peak load may be significantly affected.

## 2. Cooling load calculation methods in practice

Load calculations should accurately describe the building. All load calculation inputs should be as accurate as reasonable, without using safety factors. Introducing compounding safety factors at multiple levels in the load calculation results in an unrealistic and oversized load.

Variation in heat transmission coefficients of typical building materials and composite assemblies, differing motivations and skills of those who construct the building, unknown filtration rates, and the manner in which the building is actually operated are some of the variables that make precise calculation impossible. Even if the designer uses reasonable procedures to account for these factors, the calculation can never be more than a good estimate of the actual load. Frequently, a cooling load must be calculated before every parameter in the conditioned space can be properly or completely defined. An example is a cooling load estimate for a new building with many floors of unleased spaces for which detailed partition requirements, furnishings, lighting, and layout cannot be predefined. Potential tenant modifications once the building is occupied also must be considered. Load estimating requires proper engineering judgment that includes a thorough understanding of heat balance fundamentals.

**Perimeter spaces**[①] exposed to high solar heat gain often need cooling during sunlit portions of traditional heating months, as do completely interior spaces with significant internal heat gain. These spaces can also have significant heating loads during nonsunlit hours or after periods of nonoccupancy, when adjacent spaces have cooled below interior design temperatures. The heating loads involved can be estimated conventionally to offset or to compensate for them and prevent overheating, but they have no direct relationship to the spaces' design heating loads.

Correct design and sizing of air-conditioning systems require more than calculation of the cooling load in the space to be conditioned. The type of air-conditioning system, ventilation rate, reheat, fan energy, fan location, duct heat loss and gain, duct leakage, heat extraction lighting systems, type of return air system, and any sensible or latent heat recovery all affect system load and component sizing. Adequate system design and component sizing require that system performance be analyzed as a series of **psychrometric processes**[②].

System design could be driven by either sensible or latent load, and both need to be checked. When a space is sensible-load-driven, which is generally the case, the cooling supply air will have surplus capacity to dehumidify, but this is commonly permissible. For a space driven by latent load, (e.g., an auditorium), supply airflow based on sensible load is likely not have enough dehumidifying capability, so subcooling and reheating or some other dehumidification process is needed.

## 3. Data assembly

Calculating space cooling loads requires detailed building design information and weather data at design conditions. Generally, the following information should be compiled.

---

①② 见 Part 3 Extension.

**Building characteristics.** Building materials, component size, external surface colors, and shape are usually determined from building plans and specifications.

**Configuration.** Determine building location, orientation, and external shading from building plans and specifications. Shading from adjacent buildings can be determined from a site plan or by visiting the proposed site, but its probable permanence should be carefully evaluated before it is included in the calculation. The possibility of abnormally high ground-reflected solar radiation (e.g., from adjacent water, sand, or parking lots) or solar load from adjacent reflective buildings should not be overlooked.

**Outdoor design conditions**③. Obtain appropriate weather data, and select outdoor design conditions. ASHRAE provides information for many weather stations; note, however, that these design dry-bulb and mean coincident wet-bulb temperatures may vary considerably from data traditionally used in various areas. Use judgment to ensure that results are consistent with expectations. Also, consider prevailing wind velocity and the relationship of a project site to the selected weather station.

To estimate conductive heat gain through exterior surfaces and infiltration and outdoor air loads at any time, applicable outdoor dry- and wet-bulb temperatures must be used. There are monthly cooling load design values of outdoor conditions for many locations. These are also generally midafternoon conditions; for other times of day, the daily range profile method can be used to estimate dry- and wet-bulb temperatures. Peak cooling load is often determined by solar heat gain through fenestration; this peak may occur in winter months and/or at a time of day when outside air temperature is not at its maximum.

**Indoor design conditions.** Select indoor dry-bulb temperature, indoor relative humidity, and ventilation rate. Include permissible variations and control limits.

**Internal heat gains and operating schedules.** Obtain planned density and a proposed schedule of lighting, occupancy, internal equipment, appliances, and processes that contribute to the internal thermal load.

**Areas.** Use consistent methods for calculation of building areas. For fenestration, the definition of a component's area must be consistent with associated ratings.

*Gross surface area.* It is efficient and conservative to derive gross surface areas from outside building dimensions, ignoring wall and floor thicknesses and avoiding separate accounting of floor edge and wall corner conditions. Measure floor areas to the outside of adjacent exterior walls or to the center line of adjacent partitions. When apportioning to rooms, facade area should be divided at partition center lines. Wall height should be taken as floor-to-floor height.

*Fenestration area.* Fenestration ratings [U-factor and solar heat gain coefficient (SHGC)] are based on the entire product area, including frames. Thus, for load calculations, fenestration area is the area of the rough opening in the wall or roof.

*Net surface area.* Net surface area is the gross surface area less any enclosed fenestration area.

**Keywords:** Cooling load, heat gain, time delay effect

---

③ 见 Part 3 Extension.

**Source from:**
ASHRAE. 2009 ASHRAE Handbook. Fundamentals-Chapter 18. Atlanta, GA: ASHRAE, 2009.

# Part 3　Extension

1. Perimeter space(周边区)

对大型建筑物来说,周边区(进深6 m左右的区域)受到室外空气和日照的影响大,冬夏季空调负荷变化大。内部区由于远离外围护结构,室内负荷主要是人体、照明、设备等的发热,可能为全年冷负荷。因此,通常将平面分为周边区和内部区,对应于各区负荷变化特点分别进行空调系统设计。

2. Psychrometric processes(空气处理过程)

湿空气的状态参数主要有温度、相对湿度、含湿量、焓等。通常的湿空气特性图(psychrometric chart)是以焓和含湿量为坐标的焓湿图。借助焓湿图不仅可以查出空气的各种参数,还可以确定湿空气的各种状态变化。在北美和西欧国家使用的焓湿图是和我国不同的。图7.3 为ASHRAE 设计手册中标准大气压下的焓湿图。线图左边的分度尺示有两个刻度:一个是显热系数(SHF),即显热与全热的比值;另一个是焓差和湿度比差之比。

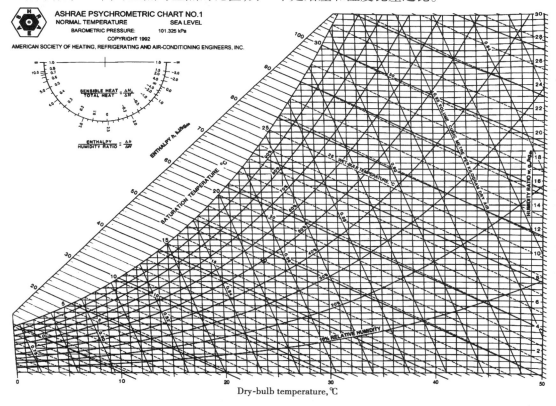

**Fig. 7.3　Psychrometric chart**

3. Outdoor Design Conditions(室外设计条件)

室外气候条件是进行建筑热环境计算分析的必备条件。在工程设计领域,为了保证建筑热环境的满意率,设计人员在进行系统和设备的设计计算时往往考虑最不利工况,因此需要具有代表性的统计气象数据。暖通空调设计用的室外气象参数就是以不保证率的统计方法为基础获得的代表性气象数据。

## Part 4　Words and Expressions

| | |
|---|---|
| building envelope | 建筑围护结构 |
| moisture migration | 水分迁移 |
| intricately | *adv.* 杂乱地；复杂地；难懂地 |
| magnitude | *n.* 量级；巨大，广大；重大，重要；（地震）级数 |
| in phase with | 与……相同 |
| the peak cooling load | 峰值冷负荷 |
| intermediate season | 过渡季节 |
| sensible heat | 显热 |
| latent heat | 潜热 |
| humidity ratio | 含湿量 |
| furniture | *n.* 家具；设备；附属品 |
| composite | *adj.* 混合成的；[建]综合式的；[数]可分解的；[植]菊科的<br>*n.* 合成物；混合物，复合材料；[植]菊科植物 |
| instantaneous heat gain | 瞬时得热 |
| heat extraction rate | 除热量 |
| intermittent operation | 间歇运行 |
| time lag | 时间延迟 |
| safety factor | 富裕系数 |
| unleased | *adj.* 未租出的 |
| predefine | *vt.* 预定义；预先确定 |
| modification | *n.* 修改，变更，改进，缓和，减轻，限制；[语]修饰，（用变音符号的）母音改变；[生]诱发变异，变体，变型 |
| adjacent space | 邻室 |
| heat extraction lighting system | 采暖照明系统 |
| auditorium | *n.* 观众席，听众席；礼堂，会堂 |
| subcool | *vt.* 使过冷，使低温冷却 |
| orientation | *n.* 方向，定位，取向，排列方向；任职培训；（外交等的）方针[态度]的确定；环境判定 |
| prevailing wind | 盛行风，主导风 |
| daily range | 日较差 |
| fenestration | *n.* 开窗法，（外科）开窗术 |
| apportion | *vt.* 分摊，分配 |
| U-factor | 传热系数 |
| solar heat gain coefficient | 太阳得热系数 |

## Part 5  Reading

## General Considerations of Energy Estimating and Modeling Methods

Energy requirements and fuel consumption of HVAC systems directly affect a building's operating cost and indirectly affect the environment. This chapter discusses methods for estimating energy use for two purposes: modeling for building and HVAC system design and associated design optimization (forward modeling), and modeling energy use of existing buildings for establishing baselines and calculating retrofit savings (data-driven modeling).

### 1. Models and approaches

A mathematical model is a description of the behavior of a system. It is made up of three components:

- Input variables, which act on the system. There are two types: controllable by the experimenter, and uncontrollable (e.g., climate).
- System structure and parameters/properties, which provide the necessary physical description of the system (e.g., thermal mass or mechanical properties of the elements).
- Output variables, which describe the reaction of the system to the input variables. Energy use is often a response variable.

The science of mathematical modeling as applied to physical systems involves determining the third component of a system when the other two components are given or specified. There are two broad but distinct approaches to modeling; which to use is dictated by the objective or purpose of the investigation.

**Forward (classical) approach.** The objective is to predict the output variables of a specified model with known structure and known parameters when subject to specified input variables. To ensure accuracy, models have tended to become increasingly complex, especially with the advent of cheap and powerful computing power. This approach presumes detailed knowledge not only of the various natural phenomena affecting system behavior but also of the magnitude of various interactions (e.g., effective thermal mass, heat and mass transfer coefficients, etc.). The main advantage of this approach is that the system need not be physically built to predict its behavior. Thus, this approach is ideal in the preliminary design and analysis stage and is most often used then.

Forward modeling of building energy use begins with a physical description of the building system or component of interest. For example, building geometry, geographical location, physical characteristics (e.g., wall material and thickness), type of equipment and operating schedules, type of HVAC system, building operating schedules, plant equipment, etc., are specified. The peak and average energy use of such a building can then be predicted or simulated by the forward simulation model. The primary benefits of this method are that it is based on sound engineering principles usually taught in colleges and universities, and consequently has gained widespread acceptance by the design and professional community. Major government-developed simulation codes, such as BLAST,

DOE-2, and Energy Plus, are based on forward simulation models. Figure 7.4 illustrates the ordering of the analysis typically performed by a building energy simulation program.

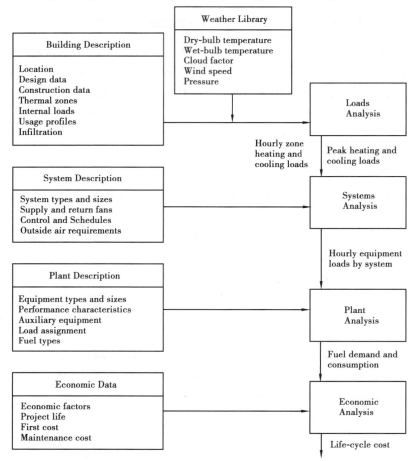

**Fig. 7.4 Flow chart for building energy simulation program**

**Data-driven (inverse) approach.** In this case, input and output variables are known and measured, and the objective is to determine a mathematical description of the system and to estimate system parameters. In contrast to the forward approach, the data-driven approach is relevant when the system has already been built and actual performance data are available for model development and/or identification. Two types of performance data can be used: nonintrusive and intrusive. Intrusive data are gathered under conditions of predetermined or planned experiments on the system to elicit system response under a wider range of system performance than would have occurred under normal system operation. These performance data allow more accurate model specification and identification. When constraints on system operation do not permit such tests to be performed, the model must be identified from nonintrusive data obtained under normal operation.

Data-driven modeling often allows identification of system models that are not only simpler to use but also are more accurate predictors of future system performance than forward models. The data-driven approach arises in many fields, such as physics, biology, engineering, and economics.

Although several monographs, textbooks, and even specialized technical journals are available in this area, the approach has not yet been widely adopted in energy-related curricula and by the building professional community.

## 2. *Characteristics of models*

### Forward models

Although procedures for estimating energy requirements vary considerably in their degree of complexity, they all have three common elements: calculation of (1) space load, (2) secondary equipment load, and (3) primary equipment energy requirements. Here, secondary refers to equipment that distributes the heating, cooling, or ventilating medium to conditioned spaces, whereas primary refers to central plant equipment that converts fuel or electric energy to heating or cooling effect.

The first step in calculating energy requirements is to determine the space load, which is the amount of energy that must be added to or extracted from a space to maintain thermal comfort. The simplest procedures assume that the energy required to maintain comfort is only a function of the outdoor dry-bulb temperature. More detailed methods consider solar effects, internal gains, heat storage in the walls and interiors, and the effects of wind on both building envelope heat transfer and infiltration.

Although energy calculations are similar to the heating and cooling load calculations used to size equipment, they are not the same. Energy calculations are based on average use and typical weather conditions rather than on maximum use and worst-case weather. Currently, the most sophisticated procedures are based on hourly profiles for climatic conditions and operational characteristics for a number of typical days of the year or on 8,760 h of operation per year.

The second step translates the space load to a load on the secondary equipment. This can be a simple estimate of duct or piping losses or gains or a complex hour-by-hour simulation of an air system, such as variable-air-volume with outdoor-air cooling. This step must include calculation of all forms of energy required by the secondary system (e.g., electrical energy to operate fans and/or pumps, as well as energy in the form of heated or chilled water).

The third step calculates the fuel and energy required by the primary equipment to meet these loads and the peak demand on the utility system. It considers equipment efficiencies and part-load characteristics. It is often necessary to keep track of the different forms of energy, such as electrical, natural gas, or oil. In some cases, where calculations are required to ensure compliance with codes or standards, these energies must be converted to source energy or resource consumed, as opposed to energy delivered to the building boundary.

Often, energy calculations lead to an economic analysis to establish the cost-effectiveness of conservation measures. Thus, thorough energy analysis provides intermediate data, such as time of energy usage and maximum demand, so that utility charges can be accurately estimated. Although not part of the energy calculations, estimated capital equipment costs should be included.

Complex and often unexpected interactions can occur between systems or between various modes of heat transfer. For example, radiant heating panels affect space loads by raising the mean radiant temperature in the space. As a result, air temperature can be lowered while maintaining comfort. Compared to a conventional heated-air system, radiant panels create a greater temperature difference from the inside surface to the outside air. Thus, conduction losses through the walls and roof increase because the inside surface temperatures are greater. At the same time, the heating load caused by infiltration or ventilation decreases because of the reduced indoor-to-outdoor-air temperature difference. The infiltration rate may also decrease because the reduced air temperature difference reduces the stack effect.

## Data-driven models

The data-driven model has to meet requirements very different from the forward model. The data-driven model can only contain a relatively small number of parameters because of the limited and often repetitive information contained in the performance data. (For example, building operation from one day to the next is fairly repetitive.) It is thus a much simpler model that contains fewer terms representative of aggregated or macroscopic parameters (e.g., overall building heat loss coefficient and time constants). Because model parameters are deduced from actual building performance, it is much more likely to accurately capture as-built system performance, thus allowing more accurate prediction of future system behavior under certain specific circumstances. Performance data collection and model formulation need to be appropriately tailored for the specific circumstance, which often requires a higher level of user skill and expertise. In general, data-driven models are less flexible than forward models in evaluating energy implications of different design and operational alternatives, and so are not a substitutes in this regard.

To better understand the uses of data-driven models, consider some of the questions that a building professional may ask about an existing building with known energy consumption:

• How does consumption compare with design predictions (and, in case of discrepancies, are they caused by anomalous weather, unintended building operation, improper operation, or other causes)?

• How would consumption change if thermostat settings, ventilation rates, or indoor lighting levels were changed?

• How much energy could be saved by retrofits to the building shell, changes to air handler operation from CV to VAV, or changes in the various control settings?

• If retrofits are implemented, can one verify that the savings are due to the retrofit and not to other causes (e.g., the weather)?

• How can one detect faults in HVAC equipment and optimize control and operation?

All these questions are better addressed by the data-driven approach. The forward approach could also be used, for example, by going back to the blueprints of the building and of the HVAC system, and repeating the analysis performed at the design stage using actual building schedules and operating modes, but this is tedious and labor-intensive, and materials and equipment often perform differently in reality than as specified. Tuning the forward-simulation model is often awkward and labor intensive, although it is still an option (as adopted in the calibrated data-driven approach).

## 3. *Choosing an analysis method*

The most important step in selecting an energy analysis method is matching method capabilities with project requirements. The method must be capable of evaluating all design options with sufficient accuracy to make correct choices. The following factors apply generally:

- **Accuracy.** The method should be sufficiently accurate to allow correct choices. Because of the many parameters involved in energy estimation, absolutely accurate energy prediction is not possible. ANSI/ASHRAE Standard 140, Method of Test for the Evaluation of Building Energy Analysis Computer Programs, was developed to identify and diagnose differences in predictions that may be caused by algorithmic differences, modeling limitations, coding errors, or input errors.
- **Sensitivity.** The method should be sensitive to the design options being considered. The difference in energy use between two choices should be adequately reflected.
- **Versatility.** The method should allow analysis of all options under consideration. When different methods must be used to consider different options, an accurate estimate of the differential energy use cannot be made.
- **Speed and cost.** The total time (gathering data, preparing input, calculations, and analysis of output) to make an analysis should be appropriate to the potential benefits gained. With greater speed, more options can be considered in a given time. The cost of analysis is largely determined by the total time of analysis.
- **Reproducibility.** The method should not allow so many vaguely defined choices that different analysts would get completely different results.
- **Ease of use.** This affects both the economics of analysis (speed) and the reproducibility of results.

## 4. *Selecting energy analysis computer programs*

Selecting a building energy analysis program depends on its application, number of times it will be used, experience of the user, and hardware available to run it. The first criterion is the capability of the program to deal with the application. For example, if the effect of a shading device is to be analyzed on a building that is also shaded by other buildings part of the time, the ability to analyze detached shading is an absolute requirement, regardless of any other factors.

Because almost all manual methods are now implemented on a computer, selection of an energy analysis method is the selection of a computer program. The cost of the computer facilities and the software itself are typically a small part of running a building energy analysis; the major costs are of learning to use the program and of using it. Major issues that influence the cost of learning a program include (1) complexity of input procedures, (2) quality of the user's manual, and (3) availability of a good support system to answer questions. As the user becomes more experienced, the cost of learning becomes less important, but the need to obtain and enter a complex set of input data continues to consume the time of even an experienced user until data are readily available in electronic form compatible with simulation programs.

Complexity of input is largely influenced by the availability of default values for the input variables. Default values can be used as a simple set of input data when detail is not needed or when building design is very conventional, but additional complexity can be supplied when needed. Secondary defaults, which can be supplied by the user, are also useful in the same way. Some programs allow the user to specify a level of detail. Then the program requests only the information appropriate to that level of detail, using default values for all others.

Quality of output is another factor to consider. Reports should be easy to read and uncluttered. Titles and headings should be unambiguous. Units should be stated explicitly. The user's manual should explain the meanings of data presented. Graphic output can be very helpful. In most cases, simple summaries of overall results are the most useful, but very detailed output is needed for certain studies and also for debugging program input during the early stages of analysis.

Before a final decision is made, manuals for the most suitable programs should be obtained and reviewed, and, if possible, demonstration versions of the programs should be obtained and run, and support from the software supplier should be tested. The availability of training should be considered when choosing a more complex program.

Availability of weather data and a weather data processing subroutine or program are major features of a program. Some programs include subroutine or supplementary programs that allow the user to create a weather file for any site for which weather data are available. Programs that do not have this capability must have weather files for various sites created by the program supplier. In that case, the available weather data and the terms on which the supplier will create new weather data files must be checked.

Auxiliary capabilities, such as economic analysis and design calculations, are a final concern in selecting a program. An economic analysis may include only the ability to calculate annual energy bills from utility rates, or it might extend to calculations or even to life-cycle cost optimization. An integrated program may save time because some input data have been entered already for other purposes.

The results of computer calculations should be accepted with caution, because the software vendor does not accept responsibility for the correctness of calculations or use of the program. Manual calculation should be done to develop a good understanding of underlying physical processes and building behavior. In addition, the user should (1) review the computer program documentation to determine what calculation procedures are used, (2) compare results with manual calculations and measured data, and (3) conduct sample tests to confirm that the program delivers acceptable results.

## Source from:
ASHRAE. 2009 ASHRAE Handbook. Fundamentals-Chapter 19. Atlanta. GA: ASHRAE, 2009.

## Part 6  Exercises and Practices

### 1. *Discussion*

(1) What factors are influencing the building cooling load?

(2) What are the objectives of building energy simulation?

### 2. *Translation*

(1) Sensible heat is added directly to the conditioned space by conduction, convection, and/or radiation. Latent heat gain occurs when moisture is added to the space (e.g., from vapor emitted by occupants and equipment).

(2) There is always significant delay between the time a heat source is activated, and the point when reradiated energy equals that being instantaneously stored.

(3) Variation in heat transmission coefficients of typical building materials and composite assemblies, differing motivations and skills of those who construct the building, unknown filtration rates, and the manner in which the building is actually operated are some of the variables that make precise calculation impossible.

(4) Perimeter spaces1 exposed to high solar heat gain often need cooling during sunlit portions of traditional heating months, as do completely interior spaces with significant internal heat gain.

(5) Peak cooling load is often determined by solar heat gain through fenestration; this peak may occur in winter months and/or at a time of day when outside air temperature is not at its maximum.

### 3. *Writing*

Please explain the difference between heat gain and cooling load in 200~300 word.

# Lesson 8
# Air Conditioning Systems

## Part 1  Preliminary

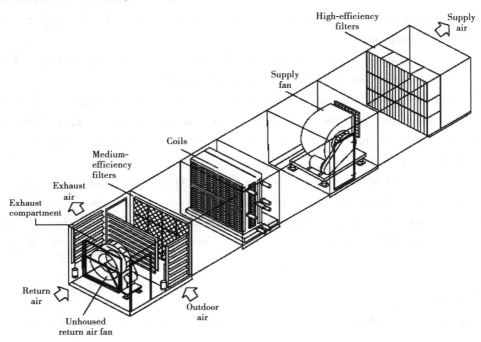

Fig. 8.1  A typical AHU

An air-handling unit (AHU) is the primary equipment in an air system of a central hydronic system. It usually consists of supply fan(s), filter(s), a cooling coil, a heating coil, a mixing box, and other accessories. In an AHU, the required amounts of outdoor air and recirculating air are often mixed and conditioned. The temperature of the discharge air is then maintained within predetermined limits by means of control systems. After that, the conditioned supply air is provided with motive force and is distributed to various conditioned spaces through ductwork and space diffusion devices.

## Part 2  Text

空调系统及设备的选择需要根据舒适性空调或工艺性空调的空气参数控制要求，综合考虑能耗特性、初投资与全生命周期成本，同时兼顾业主需求、空间限制、不同系统形式的优缺点等多方面的因素。根据散热设备向房间的散热方式不同，供暖系统可分为对流供暖和辐射供暖。空气处理设备一般包括风机、换热盘管以及过滤器各组件。制冷设备包括压缩机、冷凝器、蒸发器以及冷却塔等。供热设备包括锅炉、热泵、加热器以及太阳能设备。系统和设备的分类多种多样,设计师可根据其特点及适用条件进行合理选取。

The purpose of this chapter is to outline the criteria used in the HVAC system and equipment selection process, to describe some of the systems and equipment available, and to develop some of the underlying philosophy and background related to system selection.

## 1. *Criteria for system and equipment selection*

The problem-solving process requires some criteria that can be applied in describing and evaluating alternatives. In the selection of HVAC systems, the following criteria are used—consciously or unconsciously—because only rarely is the problem-solving process formally applied.

### Requirements of comfort or process

These requirements include temperature, always; humidity, ventilation, and pressurization, sometimes; and zoning for better control, if needed. In theory at least, the comfort requirement should have a high priority. In practice, this criterion is sometimes subordinated to first cost or to the desires of someone in authority. This is happening less often as building occupants become more sophisticated in their expectations. Process requirements are more difficult and require a thorough inquiry by the HVAC designer into the process and its needs. Until the process is fully understood, the designer cannot provide an adequate HVAC system. Most often, different parts of the process have different temperature, humidity, pressure, and cleanliness requirements; the most extreme of these can penalize the entire HVAC system.

### Energy conservation

This is usually a code requirement and not optional. State and local building codes almost invariably include requirements constraining the use of new, nonrenewable energy. Nonrenewable refers primarily to fossil-fuel sources. Renewable sources include solar power, wind, water, geothermal, waste processing, heat reclaim, and the like. The strictest codes prohibit any form of reheat (except from reclaimed or renewable sources) unless humidity control is essential. This restriction eliminates such popular systems as terminal reheat, two-deck multizone, multizone, and constant volume dual-duct systems, although the two-fan dual-duct system is still possible and the three-duct multizone system is acceptable. Most HVAC systems for process environments have opportunities for heat reclaim and other ingenious ways of conserving energy. Off-peak thermal storage systems are becoming popular for energy cost savings, although these systems may actually consume more energy than conventional systems. Thermal storage is a variation on the age-old practice of cutting and storing ice from the lake in winter, for later use in the summer.

## First cost and life-cycle cost[①]

The first cost reflects only the initial price, installed and ready to operate. The first cost ignores such factors as expected life, ease of maintenance, and even, to some extent, efficiency, although most energy codes require some minimum efficiency rating. The life-cycle cost includes all cost factors (first cost, operation, maintenance, replacement, and estimated energy use) and can be used to evaluate the total cost of the system over a period of years. A common method of comparing the life-cycle costs of two or more systems is to convert all costs to present-worth values. Typically, first cost governs in buildings being built for speculation or short-term investment. Life-cycle costs are most often used by institutional builders—schools, hospitals, government—and owners who expect to occupy the building for an indefinite extended period. Life-cycle cost analysis requires the assumption of an interest, or discount, rate and may also include anticipated inflation.

## Desires of owner, architect, or design office

Very often, someone in authority lays down guidelines which must be followed by the designer. This is particularly true for institutional owners and major retailers. Here the designer's job is to follow the criteria of the employer or the client unless it is obvious that some requirements are unsuitable in an unusual environment. Examples of such environmental conditions are extremely high or low outside-air humidity, high altitude (which affects the AHU and air-cooled condenser capacity), and contaminated outside air (which may require special filtration and treatment).

## Space limitations

Architects can influence the HVAC system selection by the space they make available in a new building. In retrofit situations, designers must work with existing space. Sometimes in existing buildings it is necessary to take additional space to provide a suitable HVAC system. For example, in adding air conditioning to a school, it is often necessary to convert a classroom to an equipment room. Rooftop systems are another alternative where space is limited, if the building structure will support such systems. In new buildings, if space is too restricted, it is desirable to discuss the implications of the space limitations in terms of equipment efficiency and maintainability with the architect. There are ways of providing a functional HVAC system in very little space, such as individual room units and rooftop units, but these systems often have a high life-cycle cost.

## Maintainability

This criterion includes equipment quality (the mean time between failures is commonly used); ease of maintenance (are high-maintenance items readily accessible in the unit?); and accessibility (Is the unit readily accessible? Is there adequate space around it for removing and replacing items?). Rooftop units may be readily accessible if an inside stair and a roof penthouse exist; but if

---

① 见 Part 3 Extension.

an outside ladder must be climbed, the adjective readily must be deleted. Many equipment rooms are easy to get to but are too small for adequate access or maintenance. This criterion is critical in the life-cycle cost analysis and in the long-term satisfaction of the building owner and occupants.

### Central plant versus distributed systems

Central plants may include only a chilled water source, both heating and chilled water, an intermediate temperature water supply for individual room heat pumps, or even a large, central air-handling system. Many buildings have no central plant. This decision is, in part, influenced by previously cited criteria and is itself a factor in the life-cycle cost analysis. In general, central plant equipment has a longer life than packaged equipment and can be operated more efficiently. The disadvantages include the cost of pumping and piping or, for the central AHU, longer duct systems and more fan horsepower. There is no simple answer to this choice. Each building must be evaluated separately.

### Simplicity and controllability

Although listed last, this is the most important criterion in terms of how the system will really work. There is an accepted truism that operators will soon reduce the HVAC system and controls to their level of understanding. This is not to criticize the operator, who may have had little or no instruction about the system. It is simply a fact of life. The designer who wants or needs to use a complex system must provide for adequate training—and retraining—for operators. The best rule is: Never add an unnecessary complication to the system or its controls.

## 2. *Options in system and equipment selection*

### Air-handling units

Air-handling units (AHUs) include factory-assembled package units and field-erected, built-up units. The common components are a fan or fans, cooling and/or heating coils, and air filters. Most units also include a mixing chamber with outside and return air connections with dampers. The size range is from small fan-coil units with as little as 100 $ft^3$/min capacity to built-up systems handling over 100,000 $ft^3$/min. When a package unit includes a cooling source, such as a refrigeration compressor and condenser, or a heating source, such as a gas-fired heater or electric heating coil, or both, then the unit is said to be self-contained. This classification includes heat pumps. Many systems for rooftop mounting are self-contained, with capacities as great as 100 tons or more of cooling and a comparable amount of heating. Some room units for wall or window installation have capacities as small as <u>0.5 or 0.75 ton</u>[②]. Split-system packages are also available, with the heating and/or cooling source section matching the fan-coil section but installed outdoors. The two sections are connected by piping. Cooling coils may use chilled water, brine, or refrigerant (direct expansion).

---

② 见 Part 3 Extension.

Heating coils may use steam or high- or low-temperature water; or "direct-fired" heating may be used, usually gas or electric resistance. Heat reclaim systems of various types are employed. Humidification equipment includes the steam grid, evaporative, and slinger/atomizer types. Dehumidification equipment includes the dehumidifying effect of most cooling coils as well as absorption-type dehumidifiers.

Thus, the designer has a wide range of equipment to choose from. Although generalizations are dangerous, some general rules may be applied, but the designer must also develop, through experience, an understanding of the best and worst choices. There are some criteria which are useful:

- Packaged equipment should be tested, rated, and certified in accordance with applicable standards.
- Minimum unit efficiencies or effectiveness should be in accordance with codes or higher.
- In general, packaged equipment has a lower first cost and a shorter life than equipment used in built-up systems. This is not always true, and comparisons must be made for the specific application.
- In general, packaged equipment is designed to be as small as possible for a given capacity. This may create problems of access for maintenance. Also the supplier should show that capacity ratings were determined for the package as assembled and not just for the separate components.
- In hotel guest rooms, motels and apartments, individual room units should be used to give occupants maximum control of their environment. Where many people share the same space, central systems are preferable, with controls which cannot be reset by occupants.
- Noise is a factor in almost any HVAC installation, yet noise is often neglected in equipment selection and installation. Noise ratings are available for all types of HVAC equipment and should be used in design and specifications.

## Radiant and convective heating and cooling

Convector radiators, using steam or hot water, are one of the oldest heating methods and are still in common use. Modern systems are more compact than the old cast-iron radiators and depend more on natural convection than on radiation.

Radiant heating by means of floor, wall, or ceiling panels is common. Hot-water piping or electric resistance heating tape is used. Maximum temperatures of the surface must be limited, and there are some control problems, particularly in floor panels, due to the mass of the panel.

Radiant cooling by means of wall or ceiling panels may also be used. Surface temperatures must be kept above the dew point; therefore, any dehumidification required must be accomplished by other means.

In modern practice, radiant and/or convective heating or cooling is usually a supplement to the air system and is used primarily to offset exterior wall, roof, and radiant floor heat gains or losses.

## Refrigeration equipment

Source cooling equipment includes refrigeration compressors of reciprocating, centrifugal, and screw types; absorption chillers using steam, hot water, or direct fuel firing; water chiller heat exchangers; condensers cooled by air, water, and evaporation; cooling towers; and evaporative coolers, including spray, slinger, and drip types.

Self-contained package AHUs typically use direct-expansion cooling with reciprocating or rotary compressors. Other AHUs may use direct expansion, chilled water, or brine cooling, with the cooling medium provided by a separate, centralized, refrigeration system. Evaporative cooling is used primarily in climates with low design ambient wet-bulb temperatures, although it may be used in almost any climate to achieve some cooling. Evaporative cooler efficiencies are highest for the spray type and lowest for the drip type. Centrifugal and screw-type compressors and absorption refrigeration are used almost entirely in large central-station water or brine chillers. Absorption refrigeration may be uneconomical unless there is an adequate source of waste heat or solar energy. Air-cooled condensers are less costly to purchase and maintain than cooling towers or evaporative condensers, but they result in higher peak condensing temperatures at design conditions and may result in lower overall efficiency in the cooling system.

The selection of the source cooling equipment is influenced primarily by the selection of the AHU equipment and systems. Often both are selected at the same time. The use of individual room units does not preclude the use of central-station chillers; this combination may be preferable in many situations. For off-peak cooling with storage, a central chilling plant is an essential item.

## Heating equipment

Source heating equipment includes central plant boilers for steam and high-, medium-, and low-temperature hot water; heat pumps, both central and unitary; direct-fired heaters; solar equipment, including solar-assisted heat pumps; and geothermal and heat reclaim. Fuels include coal, oil, natural and manufactured gas, and peat. Waste products such as refuse-derived fuel (RDF) and sawdust are also being used in limited ways. Electricity for resistance heating is not a fuel in the combustion sense but is a heat source.

Self-contained package AHUs use direct-fired heaters—usually gas or electric—or heat pumps. For other systems, some kind of central plant equipment is needed. The type of equipment and fuel used is determined on the basis of the owner's criteria, local availability and comparative cost of fuels, and, to some extent, the expertise of the designer. Large central plants for high-pressure steam or high- temperature hot water, may present safety problems, are regulated by codes and require special expertise on the part of the designer, contractor, and operator. New buildings connected to existing central plants will require the use of heat exchangers, secondary pumping or condensate return pumping, and an understanding of limitations imposed by the existing plant, such as limitations on the pressure and temperature of returned water or condensate.

**Keywords:** Comfort air conditioning system, process air conditioning system, heating and cooling source

**Source from**:

Roger W Haines, Michael E Myers. HVAC System Design Handbook-Chapter 4. New York: McGraw-Hill, 2010.

## Part 3　Extension

1. Life-cycle cost(全生命周期成本)

建筑的生命周期是指从材料与构件(含原材料的开采)、规划与设计、建造与运输、运行与维护直到拆除与处理(废弃、再循环和再利用等)的全循环过程。在进行空调系统和技术的选择时,考虑全生命周期成本对实现建筑节能减排有重要意义。

2. 0.5 or 0.75 ton (0.5 或 0.75 冷吨)

Ton(standard ton of refrigeration)是标准冷吨,1 Ton 是表示 1 吨[one short ton (2,000 lb or 907 kg]0 ℃的饱和水在 24 h 冷冻到 0 ℃(32 ℉)的冰所需要的制冷量。

常用冷热量单位换算关系(Conversion factors)

1 J = $9.478 \times 10^{-4}$ Btu = $7.376 \times 10^{-1}$ ft·lbf

1 kW·h = $3.412 \times 10^{3}$ Btu = $2.655 \times 10^{6}$ ft·lbf

1 W = 3.412 Btu/h = $1.341 \times 10^{-3}$ hp = $2.844 \times 10^{-4}$ tons

1 hp = 2 545 Btu/h

Btu(British thermal unit)是英热量单位,1 Btu 是在标准大气压下,将 1 磅纯水加热,使其温度升高 1 ℃时所需要的热量。

## Part 4　Words and Expressions

| | |
|---|---|
| subordinate | *adj.* 级别或职位较低的;次要的;附属的 |
| | *n.* 部属;部下,下级 |
| | *vt.* 使……居下位,使在次级;使服从;使从属 |
| process requirement | 工艺过程要求 |
| energy conservation | 节约能源 |
| building code | 建筑规范 |
| nonrenewable energy | 不可再生能源 |
| and the like | 等等;依次类推 |
| first cost | 初始投资 |
| present-worth values | 现值 |
| speculation | *n.* 投机活动;投机买卖;思考;推断 |
| inflation | *n.* 膨胀;通货膨胀;夸张;自命不凡 |
| lay down | 放下;规定;放弃;建造 |
| air handling unit(AHU) | 空气处理机组 |
| air-cooled condenser | 风冷冷凝器 |
| individual room unit | 分散式房间空调器 |
| rooftop unit | 屋顶式机组 |
| central plant | 集中冷热源机房 |

续表

| | |
|---|---|
| truism | n. 自明之理，老生常谈 |
| cooling and / or heating coil | 冷/热盘管 |
| mixing chamber | 静压箱 |
| fan-coil unit | 风机盘管 |
| 1ft³ | 1 立方英尺 =0.028 3 立方米 |
| compressor | n. 压缩机,压气机 |
| condenser | n. 冷凝器；（尤指汽车发动机内的）电容器 |
| self-contained | adj. 独立的；设备齐全的；沉默寡言的 |
| heat pump | 热泵 |
| radiant heating | 辐射供暖 |
| radiant cooling | 辐射供冷 |
| reciprocating refrigeration compressor | 往复式制冷压缩机 |
| centrifugal refrigeration compressor | 离心式制冷压缩机 |
| screw refrigeration compressor | 螺杆式制冷压缩机 |
| absorption chiller | 吸收式制冷机组 |
| cooling tower | 冷却塔 |
| direct-expansion cooling | 直接膨胀式冷却 |
| evaporative cooling | 蒸发冷却 |
| heat reclaim | 热回收 |
| expertise | n. 专门知识或技能；专家的意见；鉴定 |

## Part 5　Reading

### Water System：Plant-Building Loop

### 1. *System description*

Plant-building loop water systems, also called primary-secondary loop (or circuit) water systems, are the widely adopted water systems for large new and retrofit commercial HVAC&R installations in the United States today. A plant-building loop chilled water, hot water, or dual-temperature water system consists of two piping loops：

### Plant loop (primary loop)

In a plant loop, there are chiller(s)/boiler(s), circulating water pumps, diaphragm expansion tank, corresponding pipes and fittings, and control systems, as shown by loop *ABFG* in Fig. 8.2. A constant volume flow rate is maintained in the evaporator of each chiller. For a refrigeration plant equipped with multiple chillers, the chilled water volume flow rate in the plant loop will vary when a chiller and its associated chiller pump are turned on or off.

## Building loop (secondary loop)

In a building loop, there are coils, terminals, probably variable-speed water pumps, two-way control valves and control systems, and corresponding pipes, fittings, and accessories, as shown by loop $BCDEC'F$ in Fig. 8.2. The water flow in the building loop is varied as the coil load is changed from the design load to part-load.

A short common pipe, sometimes also called a bypass, connects these two loops and combines them into a plant-building loop.

## 2. Control systems

For a plant-building loop water system, there are four related specific control systems:

## Coil discharge air temperature control

A sensor and a DDC system or unit controller are used for each coil to modulate the two-way control valve and the water flow into the coil. The discharge temperature after the coil can be maintained within predetermined limits.

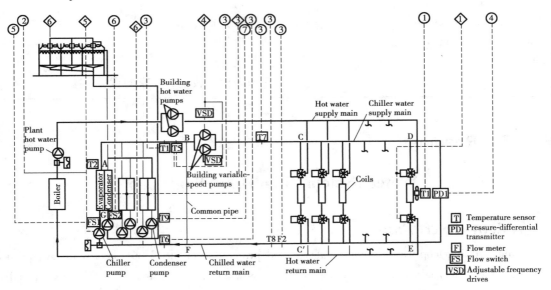

Fig. 8.2  A dual-temperature water system with plant-building loop

## Water leaving chiller temperature control

In a chiller, chilled water temperature leaving the chiller is always maintained at a preset value within a specified period by varying the refrigerant flow in the chiller. In a boiler, the leaving temperature of hot water is maintained at a predetermined value by varying the fuel flow to the burner.

During part-load, the chilled water temperature leaving the chiller is reset to a higher value, such as between 3 and 10 °F (1.7 and 5.6 ℃) according to system loads or outdoor temperature both to reduce the pressure lift between evaporating and condensing pressure and to save compressing power. ASHRAE/IESNA Standard 90.1-1999 specifies that a chilled water system with a design capacity exceeding 300,000 Btu/h (87,900 W) supplying chilled water to comfort air conditioning systems shall be equipped with controls that automatically reset supply chilled water temperature according to building loads (including return chilled water temperature) or outdoor temperature.

### Staging control

Chillers are turned on and off in sequence depending on the required system cooling capacity $Q_{sc}$ Btu/h (W), or the sum of the coils' loads. The required system cooling capacity can be found by measuring the product of the temperature difference across the supply and return mains, as shown by temperature sensors T8 and T7 and the water volume flow rate by the flowmeter F2 in Fig. 8.3. If the produced refrigeration capacity $Q_{rf}$, Btu/h (W), measured by the product of chilled water supply and return temperature differential (T6 and T5) and the flowmeter (F1) is less than $Q_{sc}$, a DDC system controller turns on a chiller. If $Q_{rf}-Q_{sc}$ is greater than the refrigeration capacity of a chiller $Q_{rfc}$, the system controller turns off a chiller. Chillers should not be staged on or off based on the chilled water volume flow rate flowing through the common pipe.

### Pressure-differential control

These controls are used to maintain the minimum required pressure differential between the supply and return mains at a specific location, as shown by PD1 in Fig. 8.3. If only one differential-pressure transmitter is installed for chilled or hot water supply and return mains, it is usually located at the end of the supply main farthest from the building pump discharge. If multiple differential-pressure transmitters are installed, they are often located at places remote from the building pump discharge, with a low signal selector to ensure that any coil in the building loop has an adequate pressure differential between the supply and return mains.

The set point of the differential-pressure transmitter should be equal to or slightly greater than the sum of the pressure drops of the control valve, coil, pipe fittings, and piping friction of the branch circuit between the supply and return mains. A low set point cannot ensure adequate water flow through the coils. A high set point consumes more pump power at a reduced flow. A set point of 15 to 25 ft (4.5 to 7.5 m) of head loss may be suitable.

## 3. *System characteristics*

For a plant-building loop chilled water system, when the volume flow rate of the chilled water in the building loop is at its design value $V_{bg,d}$, the volume flow rate in the plant loop $V_{pt}$ is equal to that in the building loop $V_{bg,d}$ theoretically, all in gpm (m³/min). In actual practice, $V_{pt}$ is slightly (less than 3 percent) higher than $V_{bg,d}$ to guarantee a sufficient chilled water supply to the building loop.

At design load, chilled water leaving the chiller(s) at point A flows through the junction of the common pipe, plant loop, and building loop (point $B$), is extracted by the variable-speed building pump; and is supplied to the coils. From the coils, chilled water returns through another junction of the building loop, common pipe, and plant loop (point $F$). There is only a very small amount of bypass chilled water in the common pipe flows in the direction from point $B$ to $F$. The chilled water return from the coils is then combined with the bypass water from the common pipe and is extracted by the chiller pump(s) and enters the chiller(s) for cooling again.

When the coils' load drops during part-load operation, and the water volume flow rate $V_{bg}$ reduces in the building loop because the control valves have been partially closed, $V_{pt}$ is now considerably greater than $V_{bg}$. Chilled water then divides into two flows at junction $B$: water at the reduced volume flow rate is extracted by the variable-speed building pump in the building loop and is supplied to the coils; the remaining water bypasses the building loop by flowing through the common pipe, is extracted by the chiller pump(s), and returns to the chiller(s).

For a water system that includes a plant-building loop with a common pipe between the two loops, Carlson (1968) states the following rule: One pumped circuit affects the operation of the other to a degree dependent on the flow and pressure drop in piping common to both circuits. The lower the pressure drop in the common pipe, the greater the degree of isolation between the plant and building loops. The head-volume flow characteristics of these loops act as two separate systems.

A plant-building loop has the following characteristics:

- It provides variable flow at the building loop with separate building pump(s) and constant flow through the evaporator of the chiller and thus saves pumping power during periods of reduced flow in the building loop. According to Rishel (1983), the annual pump energy consumption of a plant-building loop with variable flow in a building loop that uses a variable-speed building pump is about 35 percent that of a plant-through-building loop constant-flow system using three-way control valves.

- It separates the building loop from the plant loop and makes the design, operation, and control of both loops simpler and more stable.

- Based on the principles of continuity of mass and energy balance, if differences in the density of chilled water are ignored at junctions $B$ and $F$, the sum of the volume flow rates of chilled water entering the junction must be equal to the sum of volume flow rates of water leaving that junction. Also, for an adiabatic mixing process, the total enthalpy of chilled water entering the junction must be equal to the total enthalpy of water leaving the junction. At junction $B$ or $F$, chilled water has the same water pressure and temperature.

# 4. *Sequence of operations*

Consider a chilled water system in a dual-temperature water system that is in a plant-building loop, as shown in Fig. 8.3. There are three chillers in the plant loop, each of which is equipped with a constant-speed chiller pump. In the building loop, there are two variable-speed building pumps connected in parallel. One is a standby pump. Chilled water is forced through the water cooling coils in AHUs that serve various zones in the building. For simplicity, assume that the latent coil load remains constant when the coil load varies. Based on the data and information from Ellis and McKew (1996), for such a chilled water system, the sequence of operations of the DDC system is as follows:

- When the system controller of the water system is in the off position, the chiller pump is off, condenser pump is off, building pump is off, and the cooling tower fan is off.
- If the system controller is turned on, then the chiller's on / off switch in the unit controller is placed in the on position; and interlock signals are sent to three chiller pumps, one variable-speed building pump, and three condenser pumps and start all these pumps. The variable-speed building pump is always started from zero speed and increases gradually for safety and energy saving. As the chilled water flow swiches confirm that all the pumps are delivering sufficient water flow, the compressor of the leading chiller (first chiller) is turned on.
- Temperature sensor T2 tends to maintain the set point of the chilled water leaving chiller temperature often at 45 °F (7.2 °C) by means of refrigerant flow control through multiple on/off compressors, modulation of inlet vanes, or variable-speed compressor motor. Temperature sensors T7 and T8 and flowmeter F2 measure the required system cooling capacity $Q_{sc}$; and sensors T5 and T6 and flowmeter F1 measure the produced refrigeration capacity $Q_{rf}$. If $Q_{sc} > Q_{rf}$, chiller is staged on in sequence, until $Q_{rf} \geqslant Q_{sc}$.
- Condenser water temperature sensor $T$ measures the water temperature entering the condenser so that it will not be lower than a limit recommended by the manufacturer for normal operation.
- When the coils' control valves in AHUs close, the chilled water flow drops below the design flow. As the pressure-differential transmittter DP1 senses that the pressure differential between chilled water supply and return mains increases to a value which exceeds the set point, such as 15 ft (4.5 m), the system controller modulates the variable-speed drive (VSD) and reduces the speed of the variable-speed pump to maintain a 15-ft (4.5-m) pressure differential.
- At the design system load, three chillers shall provide nearly their maximum cooling capacity, and the veriable-speed building pump shall provide maximum flow through pump speed control. All two-way valves shall be nearly opened fully. A constant chilled water flow is maintained in the evaporator of each chiller. Cooling tower fan shall be continuously operated at full speed.
- During part-load operation as the sum of the coils load (system load) decreases, the two-way valves close their openings to reduce the chilled water flowing through the coils. At a specific fraction of design sensible coil load $Q_{cs}/Q_{cs,d}$, there is a corresponding water volume flow rate in the building loop, expressed as a fraction of design flow $\dot{V}_{bg}/\dot{V}_{bg,d}$, that offsets this coil load. The building variable-speed pump should operate at this building volume flow rate $\dot{V}_{bg}$ [gpm (m³/min)] with a head sufficient to overcome the head loss in the building loop through the modulation of the variable-speed pump.

The supply and return temperature differential of the building loop, or the mean chilled water temperature rise across the coils $\Delta T_{wc}$ [°F/°C] depends on the fraction of the design sensible coil load $Q_{cs}/Q_{cs,d}$, and the fraction of the design volume flow rate through the coils $\dot{V}_{bg}/\dot{V}_{bg,d}$. The smaller the value of $\dot{V}_{bg}/\dot{V}_{bg,d}$, the greater the temperature rise $\Delta T_{wc}$. At part load ($Q_{cs}/Q_{cs,d} < 1$), $\Delta T_{wc}$ is always greater than that at the design load.

- During part-load operation, temperature sensors T5, T6, T7, and T8 and flowmeters F1 and F2 measure the readings which give the produced cooling capacity $Q_{rf}$ and required system cooling capacity $Q_{sc}$. If $Q_{rf} - Q_{sc} < Q_{rf,c}$ (one chiller's cooling capacity, Btu/h), none of the chillers is staging off. When $Q_{rf} - Q_s \geq Q_{rf,c}$, one of the chillers is then turned off until $Q_{rf} - Q_{sc} < Q_{sc,c}$. To turn off a chiller, the compressor is turned off first, then the chiller pump, condenser pump, and cooling tower fans corresponding to that chiller.

- During part-load operation, a constant flow of chilled water is still maintained in the evaporator of each turned-on chiller. However, the volume flow rate in the plant loop $\dot{V}_{pt}$ depends on the number of operating chillers and their associated chiller pumps. The staging on or off of the chillers and their associated pump causes a variation of chilled water volume flow rate in the plant loop.

The difference between the volume flow rate of chilled water in the plant loop $\dot{V}_{pt}$ and the volume flow rate in the building loop $\dot{V}_{pt} - \dot{V}_{bg}$ gives the volume flow rate of chilled water in the common pipe $\dot{V}_{cn}$, that is, $\dot{V}_{cn} = \dot{V}_{pt} - \dot{V}_{bg}$. At part-load operation, there is always a bypass flow of chilled water from the plant loop returning to the chiller via the common pipe.

- During part-load operation, the set point of the chilled water leaving temperature is often reset to a higher value according to either the outdoor temperature or the reduction of system load.

- During part-load operation, as the two-way control valves close, the chilled water pressure in the supply main of the building loop tends to increase. The pressure-differential transmitter PD1 senses this increase and reduces the speed of the variable-speed building pump by means of a variable-speed drive to maintain a constant 15-ft (4.5-m) pressure differential between supply and return mains. When the system load increases, the two-way control valves open wider and PD1 senses the drop of the pressure differential, increases the speed of the pump, and still maintains a required 15-ft (4.5-m) pressure differential.

- When the water system is shut down, the system controller should be in the off position. First, the compressor(s) are turned off, then the variable-speed building pump is off, condenser pump(s) are off, chiller pump(s) are off, and cooling tower fan(s) are off. The speed of the variable-speed pump is gradually reduced to zero first, and then the pump is turned off.

**Source from:**
Shan K. Wang. Handbook of Air Conditioning and Refrigeration-Chapter 7. New York: McGraw-Hill, 2001.

## Part 6  Exercises and Practices

### 1. *Discussion*

(1) What are the classifications of air conditioning systems?

(2) Why is the variable-speed pump used?

## 2. Translation

(1) In theory at least, the comfort requirement should have a high priority. In practice, this criterion is sometimes subordinated to first cost or to the desires of someone in authority.

(2) Typically, first cost governs in buildings being built for speculation or short-term investment. Life-cycle costs are most often used by institutional builders—schools, hospitals, government—and owners who expect to occupy the building for an indefinite extended period.

(3) Noise is a factor in almost any HVAC installation, yet noise is often neglected in equipment selection and installation.

(4) In modern practice, radiant and / or convective heating or cooling is usually a supplement to the air system and is used primarily to offset exterior wall, roof, and radiant floor heat gains or losses.

(5) Evaporative cooling is used primarily in climates with low design ambient wet-bulb temperatures, although it may be used in almost any climate to achieve some cooling.

## 3. Writing

Please describe the components of air conditioning systems in 200 ~ 300 words.

# Lesson 9

# Control Systems for HVAC

## Part 1  Preliminary

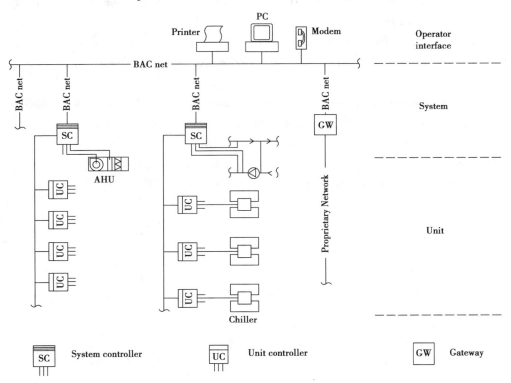

**Fig. 9.1  System architecture of a typical large EMCS**

It shows the system architecture of a typical energy management and control system with direct digital control (EMCS with DDC) for a medium or large building. Such an EMCS has mainly two operating levels: unit level and system/building level. For HVAC & R, most of the control operations are performed at the unit level. Since the software is often factory-loaded, only the time schedules, set points, and tuning constants can be changed by the user. Since a system controller has an onboard capacity, programmed by an operator or factory-preprogrammed software, to execute complicated HVAC & R and other programs, they are the brain of an EMCS.

## Part 2  Text

控制回路、执行器类型以及控制模式是暖通空调自动系统的基本要素。一个完整的控制回路包括传感器、控制器和被控设备,它的作用就是将被控变量维持在所要求的数值上或一定的范围内。根据系统控制方式有无反馈信息将自控系统分为闭环控制和开环控制。被控设备的执行器按其能源形式可分为气动、电动和液动三大类。在控制系统中,由于控制调节算法的不同,输出信号随输入信号变化的规律不同,其中常见的控制调节算法包括双位调节、分程控制、比例调节、比例积分调节、PID 调节、串级控制以及自适应控制、模糊控制等现代控制理论方法。

HVAC systems are sized to satisfy a set of design conditions, which are selected to generate a maximum load. Because these design conditions prevail during only a few hours each year, the HVAC equipment must operate most of the time at less than rated capacity. The function of the control system is to adjust the equipment capacity to match the load. Automatic control, as opposed to manual control, is preferable for both accuracy and economics; the human as a controller is not always accurate and is expensive. A properly designed, operated, and maintained automatic control system is accurate and will provide economical operation of the HVAC system. Unfortunately, not all control systems are properly designed, operated, and maintained.

## 1. Introduction

Control systems for HVAC do not operate in a vacuum. For any air conditioning application, first, it is necessary to have a building suitable for the process or comfort requirements. The best HVAC system cannot overcome inherent deficiencies in the building. Second, the HVAC system must be properly designed to satisfy the process or comfort requirements. Only when these criteria have been satisfied can a suitable control system be implemented.

All control systems operate in accordance with a few basic principles. These must be understood as background to the study of control devices and system applications.

## 2. Control loops

Fig. 9.2 illustrates a basic control loop as applied to a heating situation. The essential elements of the loop are a sensor, a **controller**[①], and a controlled device. The purpose of the system is to maintain the controlled variable at some desired value, called the set point. The process plant is controlled to provide the heat energy necessary to accomplish this. In the figure, the process plant includes the air-handling system and heating coil, the controlled variable is the temperature of the supply air, and the controlled device is the valve which controls the flow of heat energy to the coil. The sensor measures the air temperature and sends this information to the controller. In the controller, the measured temperature $T_m$ is compared with the set point $T_s$. The difference between the two is the error signal. The controller uses the error, together with one or more gain constants, to generate an output signal that is sent to the controlled device, which is thereby repositioned, if appropriate. This is a closed loop system, because the process plant response causes a change in the controlled variable, known as feedback, to which the control system can respond. If the sensed variable

---

① 见 Part 3 Extension.

is not controlled by the process plant, the control system is open loop. Alternate terminology to the open-loop or closed loop is the use of direct and indirect control. A directly controlled system causes a change in position of the controlled device to achieve the set point in the controlled variable. An indirectly controlled system uses an input which is independent of the controlled variable to position the controlled device. An example of a direct control signal is the use of a room thermostat to turn a space-heating device on and off as the room temperature varies from the set point. An indirect control signal is the use of the outside air temperature as a reference to reset the building heating water supply temperature.

Many control systems include other elements, such as switches, relays, and transducers for signal conditioning and amplification. Many HVAC systems include several separate control loops. The apparent complexity of any system can always be reduced to the essentials described above.

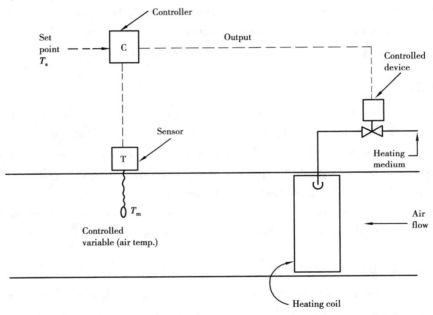

**Fig. 9.2 Elementary control loop**

## 3. Energy sources

Several types of energy are used in control systems. Most older HVAC systems use pneumatic devices, with low-pressure compressed air at 0 to 20 lb/in$^2$ gauge. Many systems are electric, using 24 to 120 V or even higher voltages. The modern trend is to use electronic devices, with low voltages and currents, for example, 0 to 10 V dc, 4 to 20 mA (milliamps), or 10 to 50 mA. Hydraulic systems are sometimes used where large forces are needed, with air or fluid pressures of 80 to 100 lb/in$^2$ or greater. Some control devices are self-contained, with the energy needed for the control output derived from the change of state of the controlled variable or from the energy in the process plant. Some systems use an electronic signal to control a pneumatic output for greater motive force.

## 4. Control modes[②]

Control systems can operate in several different modes. The simplest is the two-position mode, in which the controller output is either on or off. When applied to a valve or damper, this translates to open or closed. Fig. 9.3 illustrates two-position control. To avoid too rapid cycling, a control differential must be used. Because of the inherent time and thermal lags in the HVAC system, the operating differential is always greater than the control differential.

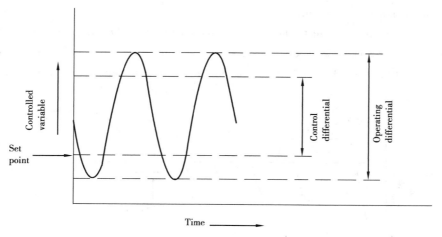

Fig. 9.3 **Two-position control**

If the output can cause the controlled device to assume any position in its range of operation, then the system is said to modulate (Fig. 9.4). In modulating control, the differential is replaced by a throttling range (sometimes called a proportional band), which is the range of controller output necessary to drive the controlled device through its full cycle (open to closed, or full speed to off).

Modulating controllers may use one mode or a combination of three modes: proportional, integral, or derivative.

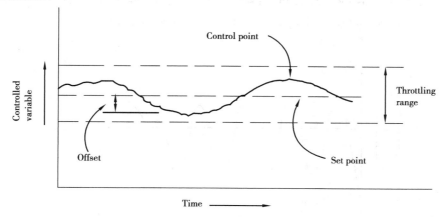

Fig. 9.4 **Modulating control**

---

② 见 Part 3 Extension.

Proportional control is common in older pneumatic control systems. This mode may be described mathematically by

$$O = A + K_p e \tag{9.1}$$

where,

$O$—controller output;

$A$—constant equal to controller output with no error signal;

$e$—error signal;

$K_p$—proportional gain constant.

The gain governs the change in the controller output per unit change in the sensor input. With proper gain control, response will be stable; i.e., when the input signal is disturbed (i.e., by a change of set point), it will level off in a short time if the load remains constant (Fig. 9.5).

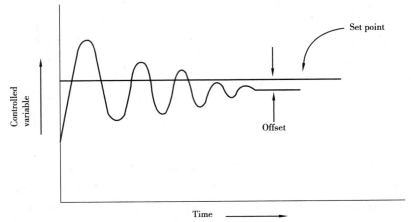

**Fig. 9.5 Proportional control, stable**

However, with proportional control, there will always be an offset—a difference between the actual value of the controlled variable and the set point. This offset will be greater at lower gains and lighter loads. If the gain is increased, the offset will be less, but too great a gain will result in instability or hunting, a continuing oscillation around the set point (Fig. 9.6).

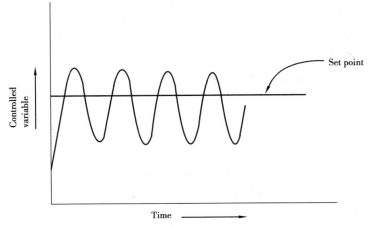

**Fig. 9.6 Proportional control, unstable**

To eliminate the offset, it is necessary to add a second term to the equation, called the integral mode:

$$O = A + K_p e + K_i \int e \, dt \tag{9.2}$$

where,

$K_i$—integral gain constant;

$\int e \, dt$ —integral of the error with respect to time.

The integral term has the effect of continuing to increase the output as long as the error persists, thereby driving the system to eliminate the error, as shown in Fig. 9.7. The integral gain $K_i$ is a function of time; the shorter the interval between samples, the greater the gain. Again, too high a gain can result in instability.

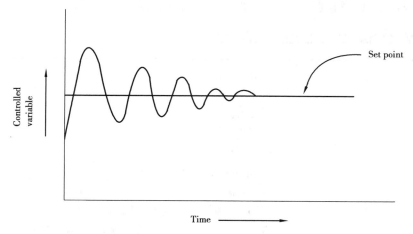

**Fig. 9.7  Proportional plus integral control**

The derivative mode is described mathematically by $K_d \, de/dt$, where $de/dt$ is the derivative of the error with respect to time. A control mode which includes all three terms is called PID (proportional-integral-derivative) mode. The derivative term describes the rate of change of the error at a point in time and therefore promotes a very rapid control response—much faster than the normal response of an HVAC system. Because of this it is usually preferable to avoid the use of derivative control with HVAC. Proportional plus integral (PI) control is preferred, and will lead to improvements in accuracy and energy consumption when compared to proportional control alone.

Most pneumatic controllers are proportional mode only, although PI mode is available. Most electronic controllers have all three modes available. In a computer-based control system, any mode can be programmed by writing the proper algorithm.

**Keywords**: Automatic control, control loop, controlled variable, controlled devices

**Source from**:

Roger W. Haines, Michael E Myers. HVAC System Design Handbook-Chapter 8. New York: McGraw-Hill, 2010.

## Part 3　Extension

1. Controller(控制器)

控制器属于自动控制系统的现场控制设备,通过读取检测装置的输入信号,按照预定的控制策略,产生输出信号,控制相关设备,从而达到控制目的。目前控制器可大致分为可编程控制器(Programmable Logic Controller,PLC)、直接数字控制器(Direct Digital Controller,DDC)、基于PC总线的工业控制计算机(工业PC),其中工业PC只用于中央站而且基本不在供暖与空调系统中使用。

2. Control model(控制模式)

控制模式即控制规律,是指在控制系统中,由于控制调节算法的不同,输出信号随输入信号变化的规律。常见的控制调节算法包括:双位调节、分程控制、比例调节、比例积分调节、PID调节、串级控制,以及自适应控制、模糊控制等现代控制理论方法。

## Part 4　Words and Expressions

| | |
|---|---|
| rated capacity | 额定能力 |
| controlled variable | 控制变量 |
| set point | 设定值 |
| error signal | 偏差信号 |
| sensed variable | 感应变量 |
| terminology | $n.$ 专有名词;术语;用词 |
| thermostat | $n.$ 恒温器 |
| switch | $n.$ 开关;转换,转换器;软鞭子;[信]接线台 |
| | $vt.$ & $vi.$ 转换;转变,改变;关闭电流;鞭打 |
| | $vt.$ 挥动(棍棒、鞭子等);迅速转动; |
| | $vi.$ 交换;调换 |
| relay | $n.$ 继电器;传递;接替人员;替班;接力赛 |
| | $vt.$ 转播,传达;用驿马递送,使接替;分程传递 |
| transducer | $n.$ 传感器,变频器,变换器 |
| amplification | $n.$ 放大 |
| pneumatic device | 气动工具 |
| hydraulic system | 液压系统 |
| two-position control | 双位控制 |
| lag | $vi.$ 走得极慢,落后 |
| | $vt.$ 落后于;给……加上外套;把……关进牢里;逮捕 |
| | $n.$ 滞后,时间间隔;囚犯;防护套;桶板 |
| modulating control | 分程控制 |
| proportional control | 比例控制 |
| level off | 平稳 |

续表

| | |
|---|---|
| offset | *n.* 偏移量;开端;出发;平版印刷;抵消 |
| | *vt.* 抵消;补偿;(为了比较的目的而)把……并列(或并置);为(管道等)装支管 |
| | *vi.* 形成分支,长出分枝;装支管 |
| | *adj.* 分支的;偏心的;抵消的;开端的 |
| oscillation | *n.* 振动;波动;动摇 |
| integral mode | 积分控制模式 |
| persist | *v.* 坚持;存留;固执;继续存在 |
| interval | *n.* 间隔;幕间休息;(数学)区间 |
| derivative mode | 微分控制模式 |

## Part 5 Reading

### Typical Control Systems

The following systems represent good basic practice. They are included to show the interaction among devices and loops in real situations. Most packaged equipment includes control systems designed and installed by the manufacturer and requiring a minimum of field installation. This is particularly true for chillers, boilers, and self-contained units.

### 1. *Control of outside air quantity*

Almost all air-handling systems utilize a mixture of outside air and recirculated air. The quantity of outside air needed is that required to replace exhausted air or that required for ventilation to maintain a desired indoor air quality (IAQ). In some industrial applications the exhaust makeup will govern, but in most commercial and institutional applications the IAQ will govern. In most places the ventilation rate will be specified by the local code authorities. The outside air volume control methods are similar for all types of air-handling systems. The essential point here is that the outside air flow must be measured and recorded to prove compliance with these codes.

Where exhaust makeup requirements govern, the outside air quantity will be fixed and controlled by a two-position damper which opens whenever the supply fan runs. This may provide 100 percent outside air or a fixed percentage of total air quantity. With 100 percent outside air it is usually recommended that the fan start-up sequence provide for opening the outside air damper before the fan starts. Low static pressure in the intake plenum may cause a collapse of that plenum.

The traditional "economy cycle" control of outside air (Fig. 9.8) operates as follows: When the supply fan is off, the outside air and relief dampers are closed. When the fan is started, these dampers open to minimum position, as determined by the setting of the minimum position switch through selector relay R2. The return air damper will be partially closed to match. If the outside air temperature is below the set point of controller C1, this controller adjust the damper positions to maintain the required minimum mixed air temperature as sensed by T2. As the outside air temperature increases, the dampers will modulate and finally reach full open for outside and relief and closed for return. As the outside air temperature continues to increase, it will reach the set point of high-limit thermostat T1 (typically about 75 °F). This will cause switch R1 to block the signal from C1, allowing the dampers to return to minimum outside air position.

The purpose of this control system is to use outside air for cooling when possible and to minimize outside air use when refrigeration is required. Note that there are three control loops involved: the main loop with T2 and C1, another loop with T1 and R1, and a third loop with the minimum position switch.

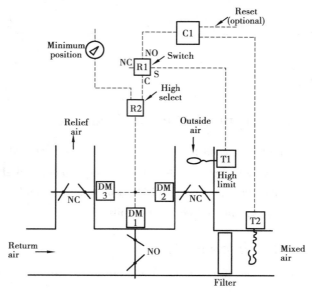

Fig. 9.8 Classical economy cycle outside-air control

With variable air volume (VAV) systems the economy cycle may no longer be satisfactory, since it provides only a fixed percentage of outside air which will result in a decrease in the quantity of outside air as the total air volume decreases. To maintain a constant minimum volume of outside air, it is necessary to use other methods. Many methods have been proposed. A few of these are:

The outside air fan uses a fixed-volume fan to force outside air into the mixed air plenum of the air-handling system. Some variations include economy cycle control as well.

Mixed air plenum pressure requires that the outside and return air dampers be controlled to provide a fixed pressure in the mixed air plenum. This pressure is assumed to be enough lower than the outside air pressure to provide the required outside air quantity. There is no provision for changes in outside pressure due to wind velocity and direction, nor for stack effect in the building, and these effects will change the outside air volume.

One hundred percent outside air with heat reclaim systems use a total heat reclaim system between outside air and exhaust air. This system is effective and economical when using modern heat reclaim equipment in mild climates.

## 2. Single-zone AHU

The single-zone air-handling unit serves one zone or a group of contiguous zones with similar loads (Fig. 9.9).

The space temperature control loop operates as follows. Controller C2 controls valves V1 (heating) and V2 (cooling) in sequence. Valve V1 is normally open and closes as the output of C2 increases. When V1 is fully closed, V2 (normally closed) begins to open; and when C2 is at maximum output, V2 will be fully open. The output of C2 depends on the supply air temperature as sensed by T3. The set point of C2 is reset from the space temperature sensor T4.

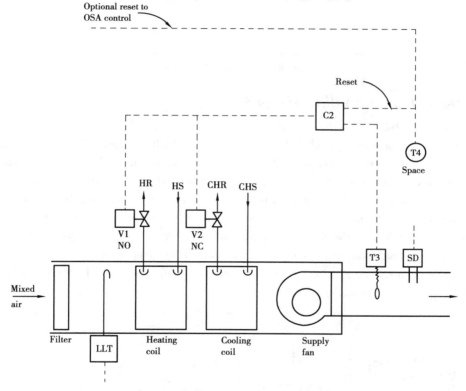

Fig. 9.9  Single-zone air-handling unit

The signal from T4 can also be used to reset the set point of the outside air controller of Fig. 9.10, for the purpose of minimizing heating energy use in mild weather.

The fire or smoke detector is required by most codes and will stop the supply fan or initiate a smoke removal cycle if fire or smoke is detected. Although the control loops are independent of one another, there is some interaction through the airstream in the AHU. Sometimes additional zoning is needed and can be provided by means of duct reheat (Fig. 9.10) or perimeter reheat by convectors, fin pipe, or baseboard radiation. Reheat is limited or prohibited by most energy codes except when used for control of humidity.

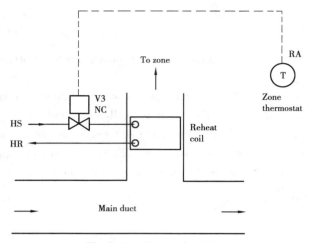

Fig. 9.10  Zone reheat coil

## 3. Dual-duct air-handling system

The dual-duct system (Fig. 9.11) operates on the same principles as the multizone system, but the hot and cold plenums are extended through the building, with a "mixing box" provided at each zone. A common arrangement (not shown) has a single fan supplying both ducts. That arrangement has the same reheat problems as the two-duct multizone system. The two-fan system shown in Fig. 9.11 overcomes most problems and provides very economical operation. The warm return air goes almost entirely to the hot duct, minimizing the use of heating energy. Outside air goes almost entirely to the cold duct, minimizing the use of cooling energy. Each fan is provided with volume control, based on maintaining a constant static pressure in its related duct. Exterior zones, which may need some heating, are provided with mixing boxes. Interior zones, which need only cooling, are provided with VAV boxes (see below).

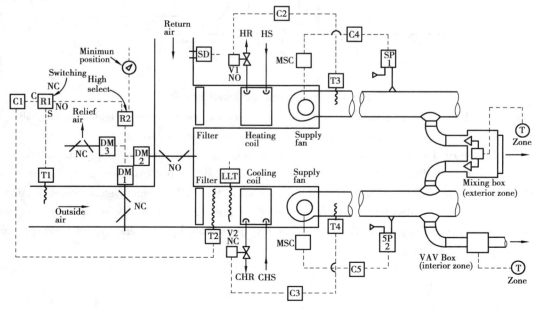

Fig. 9.11  Double-duct two-fan air-handling system

This system may be made even more economical by the use of variable volume mixing boxes (Fig. 9.12). The VAV mixing box utilizes separate operators for the hot and cold dampers, allowing them to function independently and sequentially as follows. At maximum heating, the hot damper is fully open, the cold damper closed. As the heating load decreases, the hot damper modulates toward closed, and air volume decreases. When it is about 60 percent closed, the cold damper starts to open and total air volume remains stable. The hot damper continues to close and is fully closed when the cold damper is 40 percent open. The cold damper can continue on to fully open, if required, with a corresponding increase in air volume.

When variable volume is used, the outside and return air dampers must be controlled to maintain minimum flows as described in Sec. 1.

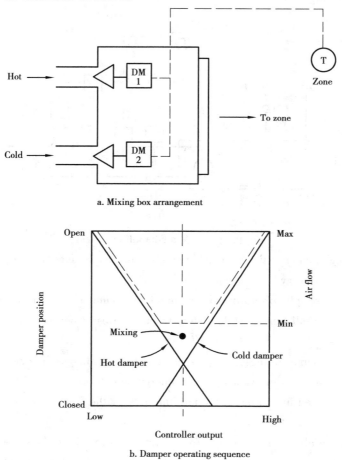

a. Mixing box arrangement

b. Damper operating sequence

**Fig. 9.12  Double-duct VAV mixing box**

## 4. *Variable-volume air-handling system*

The VAV system (Fig. 9.13) is based on the principle of matching the load by varying the air volume supplied to each zone rather than varying the temperature, with the intent of saving fan work energy as compared with a constant-volume system. As the individual VAV boxes modulate in response to zone demands, the total system volume will vary. If the fan volume is not controlled, the static pressure in the duct system will increase, resulting in noise, lack of control at the boxes, and the possibility of duct blowout. To overcome this, several methods of volume control are available. In many cases, the controlled variable is the static pressure in the duct at some point selected to provide sufficient pressure at the most remote VAV box. An old rule of thumb is to locate the sensor two-thirds to three-fourths of the distance from the fan to the most remote box. In reality, the best location is from the inlet duct to the most remote box. If this point is satisfied, all other points in the system will be satisfied. Complete building DDC systems allow every box condition to be monitored, which allows the total cfm requirements to be summed and the fan speed adjusted accordingly. Volume control methods include:

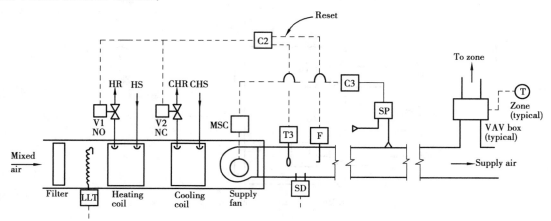

Fig. 9.13 VAV air-handling system

- A modulating damper at the fan discharge. This makes the fan "ride up its curve" and saves little, if any, fan work energy.
- A bypass from supply to return, with a modulating damper. This means that the fan is working at constant volume at all times, while system volume varies. Good control is obtained, but there are no energy savings.
- Inlet vane dampers. As these dampers modulate, they change the operating characteristic of the fan and energy is saved. Inlet vanes pose an energy penalty as an added resistance to airflow. A cone, which can be moved in and out at the fan inlet to vary the flow, has a similar effect on the inlet vane damper.
- Mechanical variable-speed drives. These systems save energy because of the physical law which states that the fan (or pump) horsepower varies as the cube of the speed. This savings is not fully realized because of mechanical losses in the systems.
- Electronic (solid-state) variable-frequency speed controllers. Variable-frequency drives are used for fans, pumps, and chiller compressors. One of the largest of such systems is used with a 5,000-hp motor driving a chiller compressor at the Dallas-Fort Worth airport. Energy use varies as the cube of the speed, although there are some losses in the electric circuits.

The single-duct VAV system provides only cooling, so supplemental heating is required in perimeter zones. This may be provided by perimeter radiation, fan-coil units, or reheat coils in the supply duct. Heating is controlled in sequence, so that it is used only when the air supply is at its minimum.

**Source from:**

Roger W. Haines, Michael E Myers. HVAC System Design Handbook-Chapter 8. New York: McGraw-Hill, 2010.

## Part 6　Exercises and Practices

### 1. *Discussion*

(1) What is the difference between the open control loop and the closed control loop?

(2) What are components of building automatic systems?

### 2. *Translation*

(1) Because these design conditions prevail during only a few hours each year, the HVAC equipment must operate most of time at less than rated capacity.

(2) This is a closed loop system, because the process plant response causes a change in the controlled variable, known as feedback, to which the control system can respond.

(3) Some control devices are self-contained, with the energy needed for the control output derived from the change of state of the controlled variable or from the energy in the process plant.

(4) In modulating control, the differential is replaced by a throttling range (sometime called a proportional band), which is the range of controller output necessary to drive the controlled device through its full cycle (open to closed, or full speed to off).

(5) The derivative term describes the rate of change of the error at a point in time and therefore promotes a very rapid control response—much faster than the normal response of an HVAC system.

### 3. *Writing*

Please give a brief introduction about control modes and their characteristics in 200 ~ 300 words.

# Lesson 10
# Ventilation System

## Part 1  Preliminary

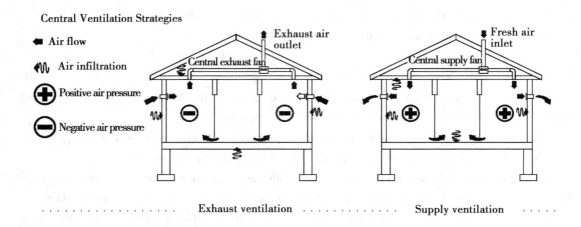

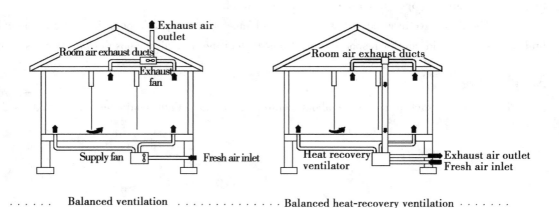

**Fig. 10.1  Ventilation strategies**

Ventilation is used to remove unpleasant smells and excessive moisture, introduce outside air, to keep interior building air circulating, and to prevent stagnation of the interior air. Ventilation includes both the exchange of air to the outside as well as circulation of air within the building. It is one of the most important factors for maintaining acceptable indoor air quality in buildings. Methods for ventilating a building may be divided into mechanical/forced and natural types.

## Part 2  Text

通风是以空气作为工作介质,采用换气方式,主要针对室内热湿环境(由温度、湿度及气流速度所表征)和室内外空气污染物浓度进行适当调控,以满足人类各种活动需求的一种建筑环境控制技术。按照通风动力的不同,通风系统可分为机械通风和自然通风两类。自然通风是依靠室外风力造成的风压和室内外空气温度差所造成的热压使空气流动;机械通风是依靠风机造成的压力使空气流动。自然通风不需要专门的动力,是一种比较经济的通风方法。

Outdoor air that flows through a building is often used to dilute and remove indoor air contaminants. However, the energy required to condition this outdoor air can be a significant portion of the total space-conditioning load. The magnitude of outdoor airflow into the building must be known for proper sizing of the HVAC equipment and evaluation of energy consumption. For buildings without mechanical cooling and dehumidification, proper ventilation and infiltration airflows are important for providing comfort for occupants. ASHRAE Standard 55 specifies conditions under which 80% or more of the occupants in a space will find it thermally acceptable.

## 1. *Ventilation and infiltration*

Air exchange of outdoor air with air already in a building can be divided into two broad classifications: ventilation and infiltration.

Ventilation is intentional introduction of air from the outside into a building; it is further subdivided into natural and mechanical ventilation. **Natural ventilation**[①] is the flow of air through open windows, doors, grilles, and other planned building envelope penetrations, and it is driven by natural and/or artificially produced pressure differentials. **Mechanical (or forced) ventilation**[②], shown in Fig. 10.2, is the intentional movement of air into and out of a building using fans and intake and exhaust vents.

Infiltration is the flow of outdoor air into a building through cracks and other unintentional openings and through the normal use of exterior doors for entrance and egress. Infiltration is also known as air leakage into a building. Exfiltration, depicted in Fig. 10.2, is leakage of indoor air out of a building through similar types of openings. Like natural ventilation, infiltration and exfiltration are driven by natural and/or artificial pressure differences. These forces are discussed in detail in the section on Driving Mechanisms for Ventilation and Infiltration. Transfer air is air that moves from one interior space to another, either intentionally or not.

Ventilation and infiltration differ significantly in how they affect energy consumption, air quality, and thermal comfort, and they can each vary with weather conditions, building operation, and use. Although one mode may be expected to dominate in a particular building, all must be considered in the proper design and operation of an HVAC system.

---

①② 见 Part 3 Extension

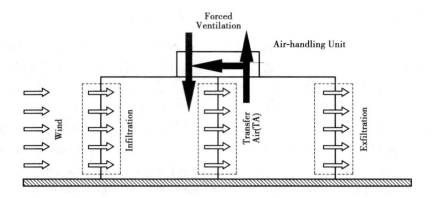

Fig. 10.2 Two-space building with mechanical ventilation, infiltration, and exfiltration

## 2. Ventilation air

Ventilation air is air used to provide acceptable indoor air quality. It may be composed of mechanical or natural ventilation, infiltration, suitably treated recirculated air, transfer air, or an appropriate combination, although the allowable means of providing ventilation air varies in standards and guidelines.

Modern commercial and institutional buildings normally have mechanical ventilation and are usually pressurized somewhat to reduce or eliminate infiltration. Mechanical ventilation has the greatest potential for control of air exchange when the system is properly designed, installed, and operated; it should provide acceptable indoor air quality and thermal comfort when ASHRAE Standard 55 and 62.1 requirements are followed.

In commercial and institutional buildings, natural ventilation (e.g., through operable windows) may not be desirable from the point of view of energy conservation and comfort. In commercial and institutional buildings with mechanical cooling and ventilation, an air- or water-side economizer may be preferable to operable windows for taking advantage of cool outdoor conditions when interior cooling is required. Infiltration may be significant in commercial and institutional buildings, especially in tall, leaky, or partially pressurized buildings and in lobby areas.

In most of the United States, residential buildings have historically relied on infiltration and natural ventilation to meet their ventilation air needs. Neither is reliable for ventilation air purposes because they depend on weather conditions, building construction, and maintenance. However, natural ventilation, usually through operable windows, is more likely to allow occupants to control airborne contaminants and interior air temperature, but it can have a substantial energy cost if used while the residence's heating or cooling equipment is operating.

In place of operable windows, small exhaust fans should be provided for localized venting in residential spaces, such as kitchens and bathrooms. Not all local building codes require that the exhaust be vented to the outside. Instead, the code may allow the air to be treated and returned to the space or to be discharged to an attic space. Poor maintenance of these treatment devices can make non-ducted vents ineffective for ventilation purposes. Condensation in attics should be avoided. In northern Europe and in Canada, some building codes require general mechanical ventilation in residences, and heat recovery heat exchangers are popular for reducing energy consumption. Low-rise residential buildings with low rates of infiltration and natural ventilation, including most new buildings, require mechanical ventilation at rates given in ASHRAE Standard 62.2.

## 3. Forced-air distribution systems

Fig. 10.3 shows a simple air-handling unit (AHU) or air handler that conditions air for a building. Air brought back to the air handler from the conditioned space is return air (RA). The return air either is discharged to the environment [exhaust air (EA)] or is reused [recirculated air (CA)]. Air brought in intentionally from the environment is outdoor or outside air (OA). Because outdoor air may need treatment to be acceptable for use in a building, it should not be called "fresh air". Outside and recirculated air are combined to form mixed air (MA), which is then conditioned and delivered to the thermal zone as supply air (SA). Any portion of the mixed air that intentionally or unintentionally circumvents conditioning is bypass air (BA). Because of the wide variety of air-handling systems, the airflows shown in Fig. 10.3 may not all be present in a particular system as defined here. Also, more complex systems may have additional airflows.

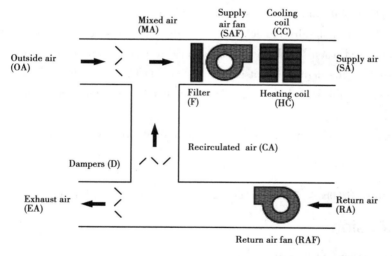

Fig. 10.3  Simple all-air air-handing unit with associated airflows

## 4. Outside air fraction

The outside airflow introduced to a building or zone by an air-handling unit can also be described by the outside air fraction $X_{oa}$, which is the ratio of the volumetric flow rate of outside air brought in by the air handler to the total supply airflow rate:

$$X_{oa} = \frac{Q_{oa}}{Q_{sa}} = \frac{Q_{oa}}{Q_{ma}} = \frac{Q_{oa}}{Q_{oa} + Q_{ca}} \qquad (10.1)$$

Where,

$X_{oa}$—the outside air fraction;

$Q_{oa}$—the outside air rate;
$Q_{sa}$—the total supply airflow rate;
$Q_{ma}$—the mixed air rate;
$Q_{ca}$—the recirculated air rate.

When expressed as a percentage, the outside air fraction is called the percent outside air. The design outside airflow rate for a building's or zone's ventilation system is found by applying the requirements of ASHRAE Standard 62.1 to that specific building. The supply airflow rate is that required to meet the thermal load. The outside air fraction and percent outside air then describe the degree of recirculation, where a low value indicates a high rate of recirculation, and a high value shows little recirculation. Conventional all-air air-handling systems for commercial and institutional buildings have approximately 10% to 40% outside air.

100% outside air means no recirculation of return air through the air-handling system. Instead, all the supply air is treated outside air, also known as makeup air (KA), and all return air is discharged directly to the outside as relief air (LA), via separate or centralized exhaust fans. An air-handling unit that provides 100% outside air to offset air that is exhausted is typically called a makeup air unit (MAU).

When outside air via mechanical ventilation is used to provide ventilation air, as is common in commercial and institutional buildings, this outside air is usually delivered to spaces as all or part of the supply air. With a variable-air-volume (VAV) system, the outside air fraction of the supply air may need to be increased when supply airflow is reduced to meet a particular thermal load. In some HVAC systems, such as the dedicated outside air system (DOAS), conditioned outside air may be delivered separately from the way the spaces' loads are handled (Mumma and Shank 2001).

**Keywords**: natural ventilation, mechanical or forced ventilation, ventilation rate

**Source from**:
ASHRAE. 2009 ASHRAE Handbook. Fundamentals-Chapter 16. Atlanta, GA: ASHRAE 2009.

# Part 3　Extension

1. Natural ventilation(自然通风)

自然通风包括风压通风和热压通风。当风吹过建筑物时,在建筑物的迎风面一侧压力升高,相对于原来大气压而言,产生了正压,在背风侧产生负压。建筑的风压作用下,在具有正值风压的一侧进风,而在负值风压的一侧排风,这就是风压作用下的自然通风。热压是由于室内外空气温度不同而形成的重力压差,这种以室内外温度差引起的压力差为动力的自然通风,称为风压作用下的自然通风,也称为"烟囱效应"。

2. Mechanical (or forced) ventilation(机械通风)

机械通风是指依靠通风机产生的压力来驱使空气流动的通风方式。机械通风是进行有组织通风的主要技术手段。机械通风由于作用压力的大小可以根据需要选择不同的风机来确定,不受自然条件的限制,因此可以通过管道把空气按要求的送风速度送至指定的任意地点,也可以从任意地点按要求的吸风速度排出被污染的空气。机械通风能适当地组织室内气流方向,并能根据需要对进风和排风进行各种处理,也便于调节通风量和稳定通风效果。但是,机械通风需要消耗电能,风机和风道等设备还会占用空间,工程设备费和维护费较大。

## Part 4　Words and Expressions

| | | |
|---|---|---|
| ventilation | *n.* | 通风 |
| airflow | *n.* | 气流 |
| infiltration | *n.* | 渗透;入渗;渗滤;下渗 |
| positive air pressure | | 正压 |
| negative air pressure | | 负压 |
| exhaust air | | 排风 |
| heat recovery | | 热回收 |
| stagnation | *n.* | 滞止,停滞 |
| outdoor(or outside) air | | 室外空气 |
| acceptable indoor air quality | | 可接受的室内空气品质 |
| contaminant | *n.* | 污染物,致污物 |
| magnitude | *n.* | 量级;巨大,广大;重大,重要;(地震)级数 |
| evaluation | *n.* | 估价;<数>赋值;估计价值;[医学]诊断 |
| energy consumption | | 能耗 |
| occupant | *n.* | 居住者;占有人 |
| air exchange | | 气体交换 |
| intentional | *adj.* | 有意的,故意的;策划的 |
| natural ventilation | | 自然通风 |
| grilles | *n.* | 格子,格栅（grille 的名词复数）;（汽车散热器的）护栅 |
| penetration | *n.* | 渗透;穿透;[军]突破;洞察力 |
| artificially | *adv.* | 人为地;人工地;不自然地;做作地 |
| pressure differential | | 压差 |
| mechanical (or forced) ventilation | | 机械通风 |
| leakage | *n.* | 泄露;漏,漏出;[商]漏损率,漏损量;渗漏物,漏出量 |
| thermal comfort | | 热舒适 |
| institutional | *adj.* | 由来已久的;习以为常的;公共机构的;慈善机构的 |
| unintentional | *adj.* | 不是故意的;无意的,无心的;无意识的 |

续表

| | |
|---|---|
| maintenance | n. 维护；维修；维持，保持；保养，保管 |
| airborne | adj. 空运的，飞机载的；航空的；空气所带的；空气传播的 |
| interior | n. 内部；内政；内地；内心<br>adj. 内部的；内地的，国内的；内面的 |
| substantial | adj. 大量的；结实的，牢固的；重大的<br>n. 本质；重要材料 |
| localized | adj. 限局性的；地区的<br>v. 使局部化；使具地方色彩 |
| condensation | n. 冷凝；冷凝液；凝结的水珠；；节略 |
| heat exchanger | 换热器，热交换器 |
| air-handling unit | 空气处理机组 |
| return air | 回风 |
| recirculated air | 循环风 |
| fresh air | 新风 |
| supply air | 送风 |
| bypass | n. 旁道，支路；迂回管道；[电]分路迂徊；[医]导管<br>vt. 疏通；忽视；管道运输 |
| additional | adj. 补充；额外的，附加的；另外的，追加的；外加 |
| outside air fraction | 新风比 |
| volumetric | adj. 测定体积的 |
| ASHRAE | abbr. American Society of Heating, Refrigerating and Air-conditioning Engineers<br>美国采暖、制冷与空调工程师学会 |
| makeup air | 补风；补充空气 |
| recirculation | n. 再通行，再流通 |
| variable-air-volume | 变风量 |

# Part 5　Reading

## Room Air Movement

### 1. *Air movement*

Air movement within spaces affects the diffusion of ventilation air and, therefore, indoor air quality and comfort. Two distinct flow patterns are commonly used to characterize air movement in rooms: displacement flow and entrainment flow. Displacement flow, shown in Fig. 10.4, is the movement of air within a space in a piston or plug-type motion. Ideally, no mixing of the room air occurs, which is desirable for removing pollutants generated within a space. A laminar-flow air distribution system that sweeps air across a space may produce displacement flow.

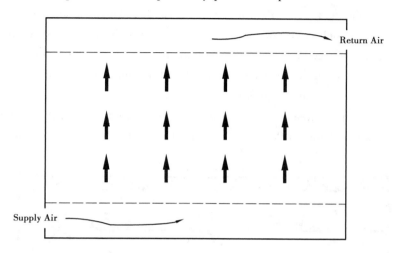

Fig. 10.4  **Displacement flow within a space**

Entrainment flow, shown in Fig. 10.5, is also known as conventional mixing. Systems with ceiling-based supply air diffusers and return air grilles are common examples of air distribution systems that produce entrainment flow. Entrainment flow with very poor mixing in the room has been called short-circuiting flow because much of the supply air leaves the room without mixing with room air. There is little evidence that properly designed, installed, and operated air distribution systems exhibit short-circuiting, although poorly designed, installed, or operated systems may short-circuit, especially ceiling-based systems in heating mode (Offermann and Int-Hout 1989).

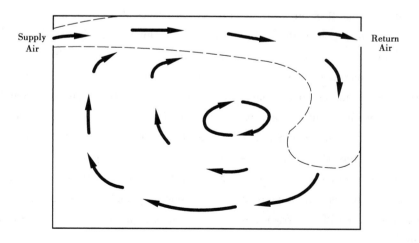

Fig. 10.5  **Entrainment flow within a space**

Perfect mixing occurs when supply air is instantly and evenly distributed throughout a space. Perfect mixing is also known as complete or uniform mixing; the air may be called well stirred or well mixed. This theoretical performance is approached by entrainment flow systems that have good mixing and by displacement flow systems that allow too much mixing (Rock et al. 1995).

Underfloor air distribution (UFAD or UAD), as shown in Fig. 10.6, is a hybrid method of conditioning and ventilating spaces (Bauman and Daly 2003). Air is introduced through a floor plenum, with or without branch ductwork or terminal units, and delivered to a space by floor-mounted diffusers. These diffusers encourage air mixing near the floor to temper the supply air. The combined air then moves vertically through the space, with reduced mixing, toward returns or exhausts placed in or near the ceiling. This vertical upward movement of the air is in the same direction as the thermal and contaminant plumes created by occupants and common equipment. Ventilation performance for UFAD systems is thus between floor-to-ceiling displacement flow and perfect mixing.

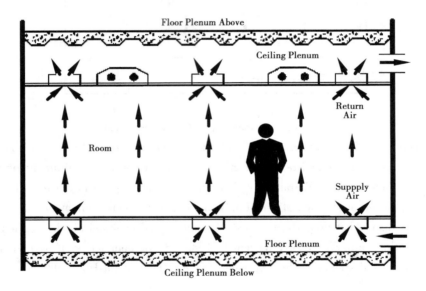

**Fig. 10.6 Underfloor air distribution to occupied space above**(Rock and Zhu 2002)

Supply air that enters a space through a diffuser is also known as primary air. A jet is formed as this primary air leaves the diffuser. Secondary air is the room air entrained into the jet. Total air is the combination of primary and secondary air at a specific point in a jet. The term primary air is also used to describe supply air provided to fan-powered mixing boxes by a central air-handling unit.

For evaluation of indoor air quality and thermal comfort, rooms are often divided into two portions: the occupied zone and the remaining volume of the space. Often, this remaining volume is solely the space above the occupants and is referred to as the ceiling zone. The occupied zone is usually defined as the lowest 1.8 m of a room, although layers near the floor and walls are sometimes deducted from it. Ceiling and floor plenums are not normally included in the occupied or ceiling zones. Thermal zones are different from these room air zones, and are defined for HVAC subsystems and their controls.

## 2. Air exchange rate

The air exchange (or change) rate $I$ compares airflow to volume and is

$$I = Q/V \qquad (10.2)$$

where,

$Q$—volumetric flow rate of air into space, $m^3/s$;

$V$—interior volume of space, $m^3$.

The air exchange rate has units of 1/time, usually $h^{-1}$. When the time unit is hours, the air exchange rate is also called air changes per hour (ach).

## 3. Time constants

Time constants $\tau$, which have units of time (usually in hours or seconds), are also used to describe ventilation and infiltration. One time constant is the time required for one air change in a building, zone, or space if ideal displacement flow existed. It is the inverse of the air exchange rate:

$$\tau = 1/I = V/Q \qquad (10.3)$$

## 4. Age of air

The age of air $\theta_{age}$ (Sandberg 1981) is the length of time $t$ that some quantity of outside air has been in a building, zone, or space. The "youngest" air is at the point where outside air enters the building by mechanical or natural ventilation or through infiltration (Grieve 1989). The "oldest" air may be at some location in the building or in the exhaust air. When the characteristics of the air distribution system are varied, age of air is inversely correlated with quality of outside air delivery. Units are of time, usually in seconds or minutes, so it is not a true efficiency or effectiveness measure. The age of air concept, however, has gained wide acceptance in Europe and is used increasingly in North America.

The age of air can be evaluated for existing buildings using tracer gas methods. Using either the decay (stepdown) or growth (stepup) tracer gas method, the zone average or nominal age of air $\theta_{age,N}$ can be determined by taking concentration measurements in the exhaust air. The local age of air $\theta_{age,L}$ is evaluated through tracer gas measurements at any desired point in a space, such as at a worker's desk. When time-dependent data of tracer gas concentration are available, the age of air can be calculated from

$$\theta_{age} = \int_{\theta=0}^{\infty} \frac{C_{in} - C}{C_{in} - C_0} d\theta \qquad (10.4)$$

where $C_{in}$ is the concentration of tracer gas being injected.

Because evaluation of the age of air requires integration to infinite time, an exponential tail is usually added to the known concentration data (Farrington et al. 1990).

## 5. Air change effectiveness

Ventilation effectiveness is a description of an air distribution system's ability to remove internally generated pollutants from a building, zone, or space. Air change effectiveness is a description of an air distribution system's ability to deliver ventilation air to a building, zone, or space. For most projects, therefore, air change effectiveness is of more relevance to HVAC system design than ventilation effectiveness. Various definitions for air change effectiveness have been proposed. The specific measure that meets local code requirements must be determined, if any is needed at all.

Air change effectiveness measures $\varepsilon_I$ are nondimensional gages of ventilation air delivery. One common definition of air change effectiveness is the ratio of a time constant to an age of air:

$$\varepsilon_I = \tau / \theta_{age} \tag{10.5}$$

An HVAC design engineer often assumes that a properly designed, installed, operated, and maintained air distribution system provides an air change effectiveness of about 1. ASHRAE Standard 129 describes a method for measuring air change effectiveness of mechanically vented spaces and buildings with limited air infiltration, exfiltration, and air leakage with surrounding indoor spaces.

**Source from:**
ASHRAE. 2009 ASHRAE Handbook. Fundamentals-Chapter 16. Atlanta, GA: ASHRAE, 2009.

## Part 6　Exercises and Practices

### 1. *Discussion*

(1) Discuss the classification of air exchange of outdoor air with air already in a building, and explain their principles and characteristics.

(2) What are the common patterns of the air movement in rooms?

### 2. *Translation*

(1) Ventilation is intentional introduction of air from the outside into a building; it is further subdivided into natural and mechanical ventilation.

(2) Natural ventilation is the flow of air through open windows, doors, grilles, and other planned building envelope penetrations, and it is driven by natural and/or artificially produced pressure differentials.

(3) Mechanical (or forced) ventilation, shown in Fig. 10.2, is the intentional movement of air into and out of a building using fans and intake and exhaust vents.

(4) Infiltration is the flow of outdoor air into a building through cracks and other unintentional openings and through the normal use of exterior doors for entrance and egress.

(5) The outside airflow introduced to a building or zone by an air-handling unit can also be described by the outside air fraction $X_{oa}$, which is the ratio of the volumetric flow rate of outside air brought in by the air handler to the total supply airflow rate.

### 3. *Writing*

Please give a brief introduction about ventilation in 200~300 words.

# Lesson 11

# Ventilation Application

## Part 1  Preliminary

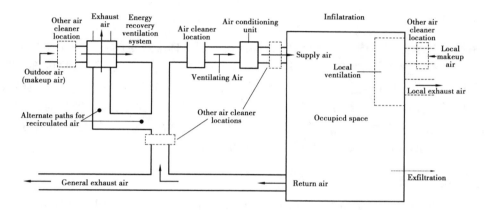

**Fig. 11.1  Ventilation system**

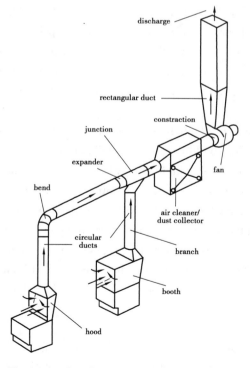

**Fig. 11.2  Local exhaust ventilation system**

Mechanical ventilation systems provide the best control and the most comfortable and uniform environment, especially when there are extremes in local climatic conditions. The systems typically consist of an inlet section, a filter, heating and/or cooling equipment, fans, ductwork, and air diffusers for distributing air within the workplace. When toxic gases or vapors are not present, air cleaned in the general exhaust system or in free-hanging filter units can be recirculated via a return duct. Air recirculation may reduce heating and cooling costs. Local exhaust ventilation systems can be the most cost-effective method of controlling air pollutants and excessive heat. Local exhaust ventilation systems typically consist of hood, ducted system, air-cleaning device, air-moving device (e.g., fan or high-pressure air ejector) and exhaust stack.

## Part 2　Text

采用通风技术控制室内空气环境,使之符合现行国家标准规定的各项卫生指标,其实质就是将室外新鲜空气(常称为"新风")或经过净化等处理的清洁空气送入建筑物内部,或是将建筑物内部的污浊空气(在经适当处理并满足规定排放标准的条件下)排至室外。通风技术的应用必须借助一定的通风系统来完成其控制室内空气环境的技术使命。进行通风设计时,应当根据环控对象的使用性质,全面考虑各种有害物的散发情况,综合运用各种通风方式及与其相应的通风系统,从而做出具有良好技术经济性能的设计方案。

### 1. *Introduction*

Traditionally, ventilation air for residences has been provided by natural ventilation and infiltration. Sherman and Matson (1997) showed that most of the older building stock is sufficiently leaky that infiltration alone can meet the minimum requirements of ASHRAE Standard 62.2. Houses built or retrofitted to new standards have substantially tighter envelopes and insufficient infiltration to meet ventilation standards. Studies have shown that concerns over safety, noise, comfort, air quality, and energy minimize occupant use of operable windows (Johnson and Long 2005; Price and Sherman 2006). As a result, these houses require supplemental mechanical ventilation to satisfy current standards.

Simply meeting minimum residential ventilation rates is not always sufficient to adequately dilute all contaminants. For some buildings, such ventilation may not meet the requirements of individuals with allergies or chemical sensitivities or when there are unusual sources such as radon or mold. In these cases, source control or extra ventilation is required to manage the contaminant levels. Therefore, especially in single-family dwellings, occupants must be responsible for introducing, monitoring, and controlling the sources in the indoor environment, as well as for operating the dwelling unit to meet their individual needs. Increasingly, residences are also used for business or hobby purposes, which may introduce air contaminants not addressed in Standard 62.2; portions of these residences may require ventilation air as required by Standard 62.1 or industrial guidelines.

### 2. *Source control*

When considering how much whole-house ventilation should be supplied, typical and unusual significant sources of indoor pollution need to be controlled. This can be done either by mitigating the source itself or by using local exhaust to extract contaminants before they can mix into the indoor environment. Typical sources that should be considered include the following:

## Clothes dryers and central vacuum systems

Clothes dryer exhaust is heavily laden with moisture and laundry by-products such as flammable lint and various gaseous contaminants. Many moisture problems have been traced to clothes dryers vented indoors. Exhaust from clothes dryers, which is typically about 70 L/s, should be vented directly to the outdoors. Similarly, central vacuum systems should be vented directly outdoors to exhaust the finer particles that pass through their filters.

## Combustion

Water and carbon dioxide are always emitted during combustion of hydrocarbons in air. Other dangerous compounds are created, as well. All these by-products should be vented directly outdoors, preferably using sealed combustion or direct-vent equipment. Venting should meet all applicable codes. For buildings with naturally aspirated combustion appliances, excessive depressurization by exhaust systems must be avoided, which can be done by keeping combustion equipment outside the pressure boundary. In addition, a depressurization safety test should be considered, such as described in ASTM Standard E1998 or CGSB Standard 51.71. Fireplace combustion products should be isolated from the occupied space using tightfitting doors and outdoor air intakes, when necessary. Flues and chimneys must be designed and installed to disperse combustion products well away from air intakes and operable windows, for example.

Carbon monoxide is one of the most pervasive indoor contaminants. It can come from virtually any source of combustion, including automobiles. Because even combustion appliances that meet manufacturers' specifications can interact with the building and emit carbon monoxide, at least one carbon monoxide alarm meeting safety standards such as CSA Standard 6.19 should be installed near sleeping areas in each dwelling, including each unit of multifamily residential buildings, that has combustion appliances (e.g., fireplaces, stoves, furnaces, water heaters) within the pressure boundary, or has attached garages or storage sheds. Carbon monoxide alarms also should be considered for nonresidential buildings: poisonings have occurred in many building types, including hotels, motels, stores, restaurants, nursing homes, dormitories, laundromats, and schools.

## Garages

Garages and storage spaces contain many sources of contaminants. Doors between them and occupied space should be well sealed with gaskets or weatherstripping and possibly be self-closing. Depressurized sections of HVAC systems, such as air handlers or return or intake ducts, should not be located in garages. If such sections must pass through garages, they must be well sealed. Care should be taken to ensure that there is a good pressure barrier between the garage and the occupied space, typically using an air/moisture retarder such as heavy polyethylene, and other measures. Carbon monoxide sources may be present in garages, so pressure barriers, fire-rated compartmentation, and ventilation of attached residences are life-safety measures. Separate ventilation systems that slightly depressurize attached garages and storage spaces and exhaust directly outdoors should be considered, especially when these support spaces are tightly constructed or are in cold climates. Several studies (Batterman et al. 2006; Emmerich et al. 2003; Fugler 2004) of contaminant sources and transport in garages found that, in some cases, significant fractions of infiltration air enter houses from attached garages, and that modern residential garages are tighter than older garages, which were commonly assumed to be leaky enough to avoid many IAQ problems.

## Particulates

The ventilation system should be designed such that return and outdoor air is filtered before passing through the thermal-conditioning components. Pressure drops associated with this filtration should be considered in the design of the air-handling system. Particulate filters or air cleaners should have a minimum efficiency of 60% for 3 μm particles, which is equivalent to a MERV 6 designated filter according to ASHRAE Standard 52.2.

## Microbiologicals

Because ventilation can increase the source as well as removal rates of various air pollutants, it is, at best, moderately effective at reducing exposures to many airborne microbiologicals. Ventilation can, however, be part of the moisture balance that is critical to retarding fungal growth on surfaces and spores released into the air, depending on indoor/outdoor conditions.

## Radon and soil gas

Buildings are exposed to gases that migrate from the soil through cracks or leaks. Soil gases vary with time and conditions, and can contain toxins from pesticides, landfill, fuel, or sewer gas, but the highest-profile pollutant in this category is radon and its radioactive-decay-produced "daughters". Source control measures, such as differential pressure control and airtightening, are far more effective than ventilation mechanisms at controlling exposure to soil gas.

## Volatile organic compounds (VOCs)

VOCs are ubiquitous in modern life. Products that emit VOCs include manufactured wood products, paints, stains, varnishes, solvents, pesticides, adhesives, wood preservatives, waxes, polishes, cleansers, lubricants, sealants, dyes, air fresheners, fuels, plastics, copy machines, printers, tobacco products, perfumes, cooking by-products, and dry-cleaned clothes. Whenever possible, VOCs and other toxic compounds should be stored outside the occupied space in loosely constructed or ventilated enclosures such as garden sheds, and away from occupied buildings' ventilation intakes. When unusual amounts of such compounds are present, additional ventilation should be considered.

## Outdoor air

Outdoor air may at times contain unacceptably high levels of pollutants, including ozone, pollen, carbon monoxide, particulate matter, odors, toxic agents, etc. At such times, it may be impossible to provide acceptable indoor air quality using solely outdoor air, and increased ventilation rates can actually decrease indoor air quality. In areas in which this problem may be anticipated, automatic or manual controls should be provided to allow reducing the ventilation rate. Cleaning recirculated air or using effective portable air cleaners should be considered for sensitive individuals.

## 3. *Local exhaust* [1]

After source elimination, the single most important source control mechanism in dwellings is local exhaust. All wet rooms and other spaces (e.g., kitchens, utility rooms, bathrooms, lavatories, toilets) designed to allow specific contaminant release should be provided with local exhaust. Workshops, recreation rooms, smoking areas, art studios, greenhouses, and hobby rooms may also require local ventilation and/or air cleaning to remove contaminants generated by the activities involved. Contaminants of concern should be evaluated to determine how much additional ventilation is required. Many of these rooms can be adequately ventilated by following the requirements for kitchens or bathrooms. If unvented combustion appliances must be used, rooms with these appliances should also meet general ventilation requirements for kitchens, because such appliances generate significant amounts of moisture and, often, ultrafine particles, even when burning properly.

Mechanical exhaust is the preferred method of providing local ventilation. Normally, it is designed to operate intermittently under manual control to exhaust contaminated air outside when the contaminant is being produced and occupants recognize the need for ventilation. However, in many circumstances, a continuous, lower-flow-rate exhaust can work as well.

Continuous Local Mechanical Exhaust. A continuously operating mechanical exhaust is intended to operate without occupant intervention. This exhaust may be part of a balanced mechanical ventilation system. The system should be designed to operate during all hours in which the dwelling is occupied. Override control should be provided if needed. The minimum delivered ventilation should be at least that given in Table 11.1.

**Table 11.1 Continuous Exhaust Airflow Rates**

| Application | Airflow Rate | Notes |
| --- | --- | --- |
| Kitchen | 5 ach | Based on kitchen volume |
| Utility room, bathroom, toilet, lavatory | 10 L/s | Not less than 2 ach |

Intermittent Local Mechanical Exhaust. An intermittently operating local mechanical exhaust is intended to be operated as needed by the occupant and should be designed with this intent. Shutoff timers, occupancy controls, multiple-speed fans, and switching integral with room lighting are helpful, provided they do not impede occupant control. The minimum airflow rate should be at least that given in Table 11.2.

---

[1] 见 Part 3 Extension.

Table 11.2  Intermittent Exhaust Airflow Rates

| Application | Airflow Rate | Notes |
|---|---|---|
| Kitchen | 50 L/s | Vented range hood required if less than 5 ach |
| Utility room, bathroom, toilet, lavatory | 25 L/s | Not less than 2 ach |

**Alternatives.** Cleaning recirculated air can sometimes be substituted for local exhaust, if it can be shown to be effective in removing contaminants of concern. Natural ventilation is not generally a suitable method for local exhaust and ventilation air needs in most climates and spaces. Using natural ventilation can cause reentrainment problems when air flows into rather than out of the space, and contaminated exhaust or exfiltrating air reenters the building. In milder climates, natural ventilation may be acceptable when the contaminant of concern is related to odor rather than health or safety. Purpose-designed passive exhaust systems have shown acceptable ventilation in some European settings, and may be considered in lieu of mechanical systems. Axley (2001b) discusses evaluation and design of passive residential ventilation systems further.

## 4. *Whole-house ventilation*[②]

Although control of significant sources of pollution in a dwelling is important, whole-house ventilation through centrally introduced, conditioned, and distributed outside air may still be needed. Each dwelling should be provided with outdoor air according to Table 11.3. The rate is the sum of the Area-Based and Occupancy-Based columns. Design occupancy can be based on the number of bedrooms as follows: first bedroom, two persons; each additional bedroom, one person. Additional ventilation should be considered when occupant densities exceed $1/25$ m$^3$.

Table 11.3  Total Ventilation Air Requirements

| Area Based | Occupancy Based |
|---|---|
| 0.15 L/s per square metre of floor space | 3.5 L/s per person, based on normal occupancy |

Natural whole-house ventilation that relies on occupant operation should not be used to make up any part of the minimum total whole-house ventilation air requirement. However, because occupancy and sources vary significantly, the capacity to ventilate above minimum rates can be provided by operable exterior openings such as doors and windows.

## 5. *Air distribution*

Ventilation air should be provided to each habitable room through mechanical and natural air distribution. If a room does not have a balance between air supply and return or exhaust, pathways for transfer air should be provided. These pathways may be door undercuts, transfer ducts with

---

② 见 Part 3 Extension.

grilles, or simply grilles where ducts are not necessary or required by code.

In houses without central air handlers, special provisions to distribute outdoor air may be required. Rooms in which occupants spend many continuous hours, such as bedrooms, may require special consideration. Local and whole-house ventilation equipment should be chosen to be energy efficient, easy to maintain, reliable, durable, and quiet. Heat recovery should be considered, especially in cold climates.

## 6. *Selection principles for residential ventilation systems*

Occupant comfort, energy efficiency, ease of use, service life, first and life-cycle cost, value-added features, and indoor environmental quality should be considered when selecting a strategy and system. HVAC and related systems can be a potential cause of poor indoor air quality. For example, occupants may not use the ventilation systems as intended if operation results in discomfort (e.g., drafts) or excessive energy use. The resulting lack of ventilation might produce poor indoor air quality. Therefore, careful design, construction, commissioning, operation, and maintenance is necessary to provide optimum effectiveness.

All exhaust, supply, or air-handler fans have the potential to change the pressure of the living space relative to the outside. High-volume fans, such as the air handler and some cooking exhaust fans, can cause high levels of depressurization, particularly in tightly constructed homes. Considering these effects is essential in design. Excessive depressurization of the living space relative to outside may cause backdrafting of combustion appliances and the migration of contaminants such as radon or other soil gases, car exhaust, or insulation particles into the living space. Depressurization can also result in moisture intrusion into building cavities in warm, moist climates, which may cause structural damage and fungal growth. Pressurization of the living space can cause condensation in building cavities in cold climates, also resulting in structural damage. Excess pressure can best be prevented by balanced ventilation systems and tightly sealed duct systems. In addition, adequate pathways must be available for all return air to the air-handling devices.

Occupant activities, operation of fans that exhaust air from the home, and leaky ducts on air conditioners, furnaces, or heat pumps may depressurize the structure. Options to address backdrafting concerns include

- Using combustion appliances with isolated (or sealed) combustion systems;
- Locating combustion appliances in a ventilated room isolated from depressurized zones by well-sealed partitions;
- Installing supply fans to balance or partially balance exhaust from the zone;
- Testing to ensure that depressurization is not excessive.

The system must be designed, built, operated, and maintained in a way that discourages growth of biological contaminants. Typical precautions include sloping condensate drain pans toward the drain, keeping condensate drains free of obstructions, keeping cooling coils free of dirt and other obstructions, maintaining humidifiers, and checking and eliminating any cause of moisture inside ducts.

Outside and exhaust airstreams of ventilation systems can be coupled using a heat pump or other device to recover thermal energy, when appropriate. Such heat pump or other equipment may reverse mode with the seasons or sensed temperature differences, for example. Heat can also be recovered from air to preheat potable water, for example.

**Keywords**: Local exhaust, Whole-house ventilation, Air distribution

**Source from**:
ASHRAE. 2009 ASHRAE Handbook. Fundamentals-Chapter 16. Atlanta, GA, ASHRAE, 2009.

## Part 3　Extension

1. Local Exhaust（局部通风）

针对建筑内部污染源集中在局部位置的情况，仅以局部污染区域为对象进行的送风或排风。局部通风与全面通风相比，通风量大大减少，环控效果亦佳，设计中应予以优先采用。

2. Whole-House Ventilation（全面通风）

针对建筑内部污染源较为分散或不确定等情况，以整个室内空间为对象进行的送风与排风，其实质是借助新风换气及稀释作用将室内的污染物加以排除，或将污染物浓度控制在卫生标准要求的范围内。

## Part 4　Words and Expressions

| | |
|---|---|
| allergy | n. [医]过敏症；[口]厌恶，反感；（对食物、花粉、虫咬等的）过敏症（allergy 的名词复数）；变态反应，变应性 |
| radon | n. <化>氡（元素符号 Rn） |
| indoor environment | 室内环境 |
| local exhaust | 局部排风 |
| central vacuum system | 中央真空吸尘系统 |
| flammable | adj. 易燃的，可燃的 |
| gaseous contaminant | 气体污染物 |
| finer particle | 细颗粒 |
| filter | n. 滤波器；滤光器；滤色镜；[化]过滤器 |
| | vi. 过滤；透过；渗透 |
| | vt. 过滤；滤除 |
| combustion | 燃烧，烧毁；氧化；骚动 |
| carbon dioxide | n. 二氧化碳 |
| hydrocarbon | n. <化>碳氢化合物，烃 |
| dangerous compounds | n. 危险化合物 |
| depressurization | 降压；泄压；弛压 |
| flue | n. 烟道 |

续表

| | |
|---|---|
| disperse | vt. & vi. （使）分散，（使）散开；散播，传布（如知识）；使（光）色散；使粒子分散 |
| | adj. 分散的 |
| carbon monoxide | n. 一氧化碳 |
| pervasive | adj. 普遍的；扩大的；渗透的；弥漫的 |
| | adv. 无处不在地；遍布地 |
| | n. 无处不在；遍布 |
| polyethylene | n. 聚乙烯 |
| particulate | adj. 微粒的，粒子的 |
| | n. 微粒，粒子 |
| pressure drops | 压力降；压降 |
| air cleaner | n. 空气过滤器；滤气器 |
| microbiological | adj. ［医］微生物学的 |
| fungal | adj. 真菌的，由真菌引起的 |
| Volatile Organic Compounds（VOCs） | 挥发性有机化合物 |
| ozone | n. ［化］臭氧；清新空气 |
| pollen | n. 花粉；［虫］粉面 |
| | vt. 传授花粉给；用花粉掩盖 |
| toxic agents | ［医］毒性药剂 |
| recirculated air | 回风，循环风 |
| continuous local mechanical exhaust | 连续性局部机械排风 |
| intermittent local mechanical exhaust | 间歇性局部机械排风 |
| whole-house ventilation | 全面通风 |
| air distribution | 风量分配 |

## Part 5　Reading

### Local Exhaust Fund Amentals

Industrial exhaust ventilation systems collect and remove airborne contaminants consisting of particulate matter (dust, fumes, smokes, and fibers), vapors, and gases that can create a hazardous, unhealthy, or undesirable atmosphere. Exhaust systems can also salvage usable material, improve plant housekeeping, and capture and remove excessive heat or moisture. Often, industrial ventilation exhaust systems are considered life-safety systems and can contain hazardous gases and/or particles.

Local exhaust ventilation systems can be the most cost-effective method of controlling air pollutants and excessive heat. For many manual operations, capturing pollutants at or near their source is the only way to ensure compliance with threshold limit values (TLVs) in the worker's breathing zone. Local exhaust ventilation optimizes ventilation exhaust airflow, thus optimizing system operating costs.

## 1. *System components*

Local exhaust ventilation systems typically consist of the following basic elements:
- Hood to capture pollutants and/or excessive heat;
- Ducted system to transport polluted air to air cleaning device or building exhaust;
- Air-cleaning device to remove captured pollutants from the airstream for recycling or disposal;
- Air-moving device (e. g. , fan or high-pressure air ejector), which provides motive power to overcome system resistance;
- Exhaust stack, which discharges system air to the atmosphere.

## 2. *System classification*

### Contaminant source type

Knowledge of the process or operation is essential before a local exhaust hood system can be designed.

### Hood type

Exhaust hoods are typically round, rectangular, or slotted to accommodate the geometry of the source. Hoods are either enclosing or nonenclosing (Fig. 11.3). Enclosing hoods provide more effective and economical contaminant control because their exhaust rates and the effects of room air currents are minimal compared to those for nonenclosing hoods. Hood access openings for inspection and maintenance should be as small as possible and out of the natural path of the contaminant. Hood performance (i. e. , how well it captures the contaminant) must be verified by an industrial hygienist.

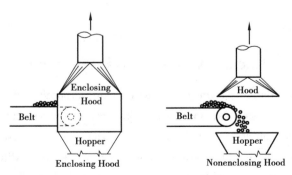

Fig. 11.3 **Enclosing and nonenclosing hoods**

A nonenclosing hood can be used if access requirements make it necessary to leave all or part of the process open. Careful attention must be paid to airflow patterns and capture velocities around the process and hood and to the process characteristics to make nonenclosing hoods effective. The more of the process that can be enclosed, the less exhaust airflow required to control the contaminant(s).

## System mobility

Local exhaust systems with nonenclosing hoods can be stationary (i.e., having a fixed hood position), moveable, portable, or built-in (into the process equipment). Moveable hoods are used when process equipment must be accessed for repair and loading and unloading of materials (e.g., in electric ovens for melting steel).

The portable exhaust system shown in Fig. 11.4 is commonly used for temporary exhausting of fumes and solvents in confined spaces or during maintenance. It has a built-in fan and filter and a linear or round exhaust hood connected to a flexible hose. Built-in local exhausts are commonly used to evacuate welding fumes, such as hoods built into stationary or turnover welding tables. Lateral exhaust hoods, which exhaust air through slots on the periphery of open vessels, such as those used for galvanizing metals, are examples of built-in local exhausts.

Fig. 11.4 Portable fume extractor with built-in fan and filter

## 3. Effectiveness of local exhaust

The most effective hood uses the minimum exhaust airflow rate to provide maximum contaminant control. Capture effectiveness should be high, but it can be difficult and costly to develop a hood that is 100% efficient. Makeup air supplied by general ventilation to replace exhausted air can dilute contaminants that are not captured by the hood. The more of the process that can be enclosed, the less exhaust airflow required to control the contaminant(s).

## Capture velocity

Capture velocity is the air velocity at the point of contaminant generation upstream of a hood. The contaminant enters the moving airstream at the point of generation and is carried along with the air into the hood. Designers use a capture velocity $V_c$ to select a volumetric flow rate to draw air into the hood. Table 11.4 shows ranges of capture velocities for several industrial operations. These figures are based on successful experience under ideal conditions. Once capture velocity upstream of the hood and hood position relative to the source are known, then the hood flow rate can be determined for the particular hood design. Velocity distributions for specific hoods must be known or determined.

**Table 11.4  Range of capture (control) velocities**

| Condition of Contaminant Dispersion | Examples | Capture Velocity, m/s |
|---|---|---|
| Released with essentially no velocity into still air | Evaporation from tanks, degreasing, plating | 0.25 to 0.5 |
| Released at low velocity into moderately still air | Container filling, low-speed conveyor transfers, welding | 0.5 to 1.0 |
| Active generation into zone of rapid air motion | Barrel filling, chute loading of conveyors, crushing, cool shakeout | 1.0 to 2.5 |
| Released at high velocity into zone of very rapid air motion | Grinding, abrasive blasting, tumbling, hot shakeout | 2.5 to 10 |

*Note*: In each category above, a range of capture velocities is shown. The proper choice of values depends on several factors (Alden and Kane 1982):

| Lower End of Range | Upper End of Range |
|---|---|
| 1. Room air currents favorable to capture | 1. Distributing room air currents |
| 2. Contaminants of low toxicity or of nuisance value only | 2. Contaminants of high toxicity |
| 3. Intermitten, low production | 3. High production, heavy use |
| 4. Large hood; large air mass in motion | 4. Small hood; local control only |

## Hood volumetric flow rate

For a given hood configuration and capture velocity, the exhaust volumetric flow rate (the airflow rate that allows contaminant capture) can be calculated as

$$Q_0 = V_0 A_a \tag{11.1}$$

where,

$Q_0$—exhaust volumetric flow rate, $m^3/s$;

$V_0$—average air velocity in hood opening that ensures capture velocity at point of contaminant release, m/s;

$A_o$—hood opening area, $m^2$.

## 4. *Principles of hood design optimization*

Numerous studies of local exhaust and common practices have led to the following hood design principles:

• Hood location should be as close as possible to the source of contamination;

• The hood opening should be positioned so that it causes the contaminant to deviate the least from its natural path;

• The hood should be located so that the contaminant is drawn away from the operator's breathing zone;

• Hood size must be the same as or larger than the cross section of flow entering the hood. If the hood is smaller than the flow, a higher volumetric flow rate is required;

• Worker position with relation to contaminant source, hood design, and airflow path should be evaluated;

• Canopy hoods (Fig. 11.5) should not be used where the operator must bend over a tank or process.

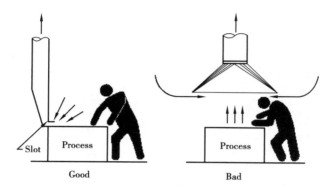

Fig. 11.5　**Influence of hood location on contamination of air in the operator's breathing zone**

**Source from:**
ASHRAE. 2007ASHRAE Handbook. HVAC Applications-Chapter 30. Atlanta, GA: ASHRAE, 2007.

## Part 6　Exercises and Practices

### 1. Discussion

(1) What are the selection principles for residential ventilation systems?

(2) Compare the local exhaust with the whole-house ventilation in a dwelling.

### 2. Translation

(1) When considering how much whole-house ventilation should be supplied, typical and unusual significant sources of indoor pollution need to be controlled.

(2) The ventilation system should be designed such that return and outdoor air is filtered before passing through the thermal-conditioning components.

(3) Mechanical exhaust is the preferred method of providing local ventilation. Normally, it is designed to operate intermittently under manual control to exhaust contaminated air outside when the contaminant is being produced and occupants recognize the need for ventilation.

(4) Although control of significant sources of pollution in a dwelling is important, whole-house ventilation through centrally introduced, conditioned, and distributed outside air may still be needed.

(5) Natural whole-house ventilation that relies on occupant operation should not be used to make up any part of the minimum total whole-house ventilation air requirement.

## 3. *Writing*

Please describe the component of local exhaust system in 200~300 words.

# Lesson 12

# An Introduction to Cleanroom

## Part 1　Preliminary

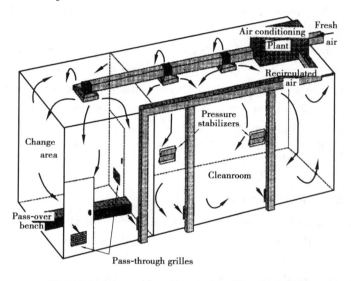

**Fig. 12.1　Conventionally ventilated cleanroom**

A cleanroom is a facility ordinarily utilized as a part of specialized industrial production or scientific research, including the manufacture of pharmaceutical items and microprocessors. Cleanrooms are designed to maintain extremely low levels of particulates, such as dust, airborne organisms, or vaporized particles. Cleanrooms typically have a cleanliness level quantified by the number of particles per cubic meter at a predetermined molecule measure. The ambient outdoor air in a typical urban area contains 35,000,000 particles for each cubic meter in the size range 0.5 μm and bigger in measurement, equivalent to an ISO 9 cleanroom, while by comparison an ISO 1 cleanroom permits no particles in that size range and just 12 particles for each cubic meter of 0.3 μm and smaller.

## Part 2　Text

洁净室是指用于专业的工艺生产或科学研究的设施,如医院、药品生产和微处理器制造等。通常根据需要对空气洁净度、温度、湿度、压力、噪声等参数控制在某一范围之内。洁净室的设计目的,是维持室内某种颗粒物含量在极低的水平,如灰尘、空气传播的有机物或挥发性颗粒物等。洁净室的洁净等级是以每立方米中某种粒径的悬浮粒子浓度限值来定义的。对于典型城市地区的室外环境而言,空气中粒径≥0.5 μm 的悬浮粒子浓度约为 3 500 000 000 pc/m³,相当于国际标准中 ISO 9 级洁净室;而 ISO 1 级洁净室则不允许该粒径范围内的颗粒存在,仅允许≤0.3 μm 的悬浮粒子存在且浓度限值为 12 pc/m³。

## 1. What is a cleanroom?

It is clear that a cleanroom is a room that is clean. However, a cleanroom now has a special meaning and it is defined in the International Organization for Standarization (ISO) standard 14644-1 as:

A room in which the concentration of airborne particles is controlled, and which is constructed and used in a manner to minimize the introduction, generation, and retention of particles inside the room and in which other relevant parameters, e. g. temperature, humidity, and pressure, are controlled as necessary.

The first two thirds of the definition is, in essence, what a cleanroom is. It is a room that minimises the introduction, generation and retention of particles. This is achieved, firstly, by supplying it with exceptionally large quantities of air that has been filtered with high efficiency filters. This air is used to (1) dilute and remove the particles and bacteria dispersed from personnel and machinery within the room and, (2) to pressurize the room and ensure that no dirty air flows into the cleanroom. Secondly, a cleanroom is built with materials that do not generate particles and can be easily cleaned. Finally, cleanroom personnel use clothing that envelops them and minimizes their dispersion of particles and micro-organisms. Cleanrooms can also control the temperature, humidity, sound, lighting, and vibration.

## 2. The need for cleanrooms

The cleanroom is a modern phenomenon. Although the roots of cleanroom design and management go back more than 100 years and are rooted in the control of infection in hospitals, the need for a clean environment for industrial manufacturing is a requirement of modern society. Cleanrooms are needed because people, production machinery and the building structure generate contamination. People and machinery produce millions of particles, and conventional building materials can easily break up. A cleanroom controls this dispersion and allows manufacturing to be carried out in a clean environment.

The use of cleanrooms is diverse and shown in Table 12.1 is a selection of products that are now being made in cleanrooms.

Table 12.1  Some clean and containment room applications

| | |
|---|---|
| Electronics | computers, TV tubes, flat screens, magnetic tape production |
| Semiconductors | production of integrated circuits used in computer memory and control |
| Micromechanics | gyroscopes, miniature bearings, compact disc players |
| Optics | lenses, photographic film, laser equipment |
| Biotechnology | antibiotic production, genetic engineering |
| Pharmacy | sterile pharmaceuticals |
| Medical devices | heart valves, cardiac by-pass systems |
| Food and drink | disease-free food and drink |
| Hospital | immunodeficiency therapy, isolation of contagious patients, operating rooms |

It may be seen in Table 12.1 that cleanroom applications can be broadly divided into two. In the top section of Table 12.1 are those industries where dust particles are a problem, and their presence, even in sub-micrometer size, may prevent a product functioning, or reduce its useful life.

A major user of cleanrooms is the semiconductor fabrication industry, where processors are produced for use in computers, cars and other machines. Fig. 12.2 shows a photomicrograph of a semiconductor with a particle on it. Such particles can cause an electrical short and ruin the semiconductor. To minimize contamination problems, semiconductors are manufactured in cleanrooms with very high standards of cleanliness.

It may also be seen from Table 12.1 that many of the examples are recent innovations and this list will certainly be added to in the future, there being a considerable and expanding demand for these types of rooms.

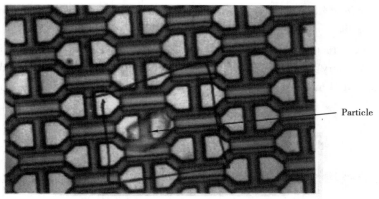

Fig. 12.2  **Contaminating particles on a semiconductor**

## 3. *Types of cleanroom*

Cleanrooms are classified by the cleanliness of their air. The method most easily understood and universally applied is the one suggested in versions of Federal Standard 209 up to edition "D" in which the number of particles equal to and greater than 0.5 wm is measured in one cubic foot of air and this count is used to classify the room. This Federal Standard has now been superseded by a metric version (Federal Standard 209E) which was published in 1992.

It should be appreciated that the airborne contamination level of cleanroom is dependent on the particle-generating activities going on in the room. If a room is empty, very low particle concentrations can be achieved, these closely reflecting the quality of air supplied and hence the removal efficiency of the high efficiency filters. If the room has production equipment in it and operating, there will be a greater particle concentration but the greatest concentration will occur when the room is in full production.

A classification of the room may therefore be carried out when the room is:

● **As built**[①]: condition where the installation is complete with all services connected and functioning but with no production equipment, materials, or personnel present.

● **At rest**: condition where the installation is complete with equipment installed and operating in a manner agreed upon by the customer and supplier, but with no personnel present.

---

① 见 Part 3 Extension.

● **Operational**: condition where the installation is functioning in the specified manner, with the specified number of personnel present and working in the manner agreed upon.

Cleanrooms have evolved into two major types and they are differentiated by their method of ventilation. These are turbulently ventilated and **unidirectional flow**[②]. cleanrooms. Turbulently ventilated cleanrooms are also known as "**nonunidirectional**"[③]. Unidirectional flow cleanrooms were originally known as "laminar flow" cleanrooms. The unidirectional type of cleanroom uses very much more air than the turbulently ventilated type, and gives superior cleanliness.

The two major types of cleanroom are shown diagrammatically in Fig. 12.3 and Fig. 12.4. Fig. 12.3 shows a turbulently ventilated room receiving clean filtered air through air diffusers in the ceiling. This air mixes with the room air and removes airborne contamination through air extracts at the bottom of the walls.

The air changes are normally equal to or greater than, 20 per hour, this being much greater than that used in ordinary rooms, such as in offices. In this style of cleanroom, the contamination generated by people and machinery is mixed and diluted with the supply air and then removed.

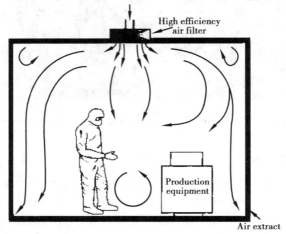

**Fig. 12.3  Conventionally ventilated type of cleanroom**

Fig. 12.4 shows the basic principles of a unidirectional flow room. High efficiency filters are installed across a whole ceiling (or wall in some systems) and these supply air. This air sweeps across the room in a unidirectional way at a speed of around 0.4 m/s and exits through the floor, thus removing the airborne contamination from the room. This system uses much more air than the turbulently ventilated cleanroom but, because of the directed air movement, it minimizes the spread of contamination about the room and sweeps it out through the floor.

Clean air devices, such as unidirectional benches or isolators, are used in both turbulently and unidirectional ventilated cleanrooms. These machines will give a localised supply of filtered air and enhanced air conditions where required, e.g. at the area where the product is open to contamination.

## 4. *What is cleanroom technology*?

As can be seen in Fig. 12.5, cleanroom technology can be divided into three broad areas.

---

②③ 见 Part 3 Extension.

These areas can also be seen to parallel the use of the technology as the cleanroom user moves from firstly deciding to purchase a room to finally operating it.

Firstly, it is necessary to design and construct the room. To do this one must consider (1) the design standards that should be used, (2) what design layout and construction materials can be used, and (3) how services should be supplied to the cleanroom.

Secondly, after the cleanroom has been installed and working, it must be tested to check that it conforms to the stipulated design. During the life of the cleanroom, the room must also be monitored to ensure that it continually achieves the standards that are required.

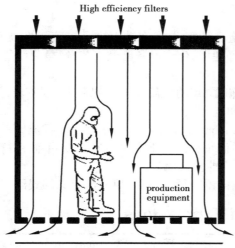

Fig. 12.4  **Unidirectional flow type of cleanroom**

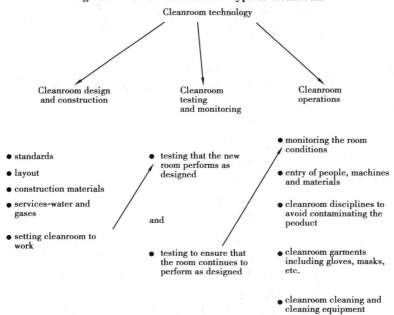

Fig. 12.5  **Various parts of cleanroom technology and their interconnections**

Finally, it is necessary to operate the cleanroom correctly so that the manufactured products are not contaminated. This requires that the entry of people and materials, the garment selection, cleanroom disciplines and the cleaning of the room are all correctly carried out.

**Keywords**: Cleanroom, Unidirectional flow, Nonunidirectional flow

**Source from**:
W. Whyte. Cleanroom Technology-Fundamentals of Design, Testing and Operation. chapter 1 Introduction. England: John Wiley & Sons Ltd, 2001.

## Part 3　Extension

1. As built(空态):设施已经建成,所有动力接通并运行,但无生产设备、物料及人员。
   At rest(静态):设施已经建成,生产设备已经安装,经净化设备生厂家认可,在无人员的状态下运行。
   Operational(动态):设施以规定的状态运行,有规定的人员在场,并在商定的状态下进行工作。
2. Unidirectional flow(单向流)
   单向流又称为层流(Laminar flow),沿单一方向呈平行流线并且与气流方向垂直的断面上风速均匀的气流。与水平面垂直的叫垂直单向流,与水平面平行的叫水平单向流。
3. Nonunidirectional flow(非单向流)
   具有多个通路或气流方向不平行,不满足单向流定义的气流。

## Part 4　Words and Expressions

| | |
|---|---|
| particles | n. 颗粒;微粒(particle 的名词复数) |
| dilute | vt. 稀释,冲淡; |
| | adj. 稀释的,冲淡的 |
| vibration | n. 震动;摆动 |
| micro-organism | n. 微生物; |
| airborne contaminants | 空气污染物 |
| semiconductors | n. 半导体(semiconductor 的名词复数) |
| micromechanics | n. 微观力学 |
| optics | n. 光学 |
| biotechnology | n. 生物技术 |
| unidirectional flow | 单向流 |
| nonunidirectional flow | 非单向流 |
| laminar flow cleanroom | 层流洁净室 |
| air diffusers | 散流器 |
| air extracts | 排风口 |
| high efficiency filters | 高效过滤器 |
| sweep | vt. 扫除;打扫,清理;彻底搜索;掠过 |
| stipulated | vt. 规定;约定; |
| | adj.〔法〕合同规定的 |
| garment | n. 服装,衣服; |
| | v. 给……穿衣服 |

## Part 5  Reading

## Design of Undirectional Cleanroom and Clean Air Devices

Cleanrooms that are ventilated in the turbulent manner may achieve conditions as low as ISO Class 6 (Class 1,000) during manufacturing, but this is more likely to be ISO Class 7 (Class 10,000). To obtain rooms better than ISO Class 6 (Class 1,000) during operation, greater dilution of the generated particles is required. This can be achieved by a unidirectional flow of air.

### 1. *Unidirectional cleanrooms*

Unidirectional airflow is used in cleanrooms when a low airborne concentration of particles or micro-organisms is required. This type of cleanroom was previously known as "laminar flow", both names describing the flow of air. The airflow is in one direction, either horizontal or vertical, at a uniform speed that is normally between 0.3 and 0.5 m/s (60 ft/min to 100 ft/min) and throughout the entire air space. Fig. 12.6 is a cross-section through a typical vertical flow type of cleanroom. It may be seen that air is supplied from a complete bank of high efficiency filters in the roof of the cleanroom. The air then flows down through the room like an air piston, thus removing the contamination. It then exits through the floor, mixes with some fresh air brought in from outside, and recirculates back to the high efficiency air filters.

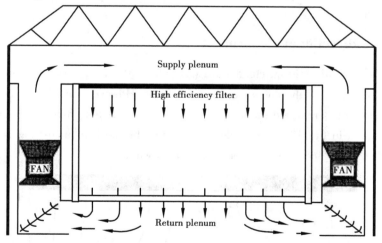

**Fig. 12.6  Vertical unidirectional flow cleanroom**

Airborne contamination from people and processes can be immediately removed by this flow of air, whereas the turbulently ventilated system relies on mixing and dilution. In an empty room with no obstructions to the unidirectional airflow, contamination can be quickly removed by air velocities much lower than those mentioned above. However, in an operating room there are machines causing obstructions to the airflow, and people moving about it. Obstructions may cause the unidirectional flow to be turned into turbulent flow and local air recirculation to be established round the obstructions. Movement of people will also turn the unidirectional flow into turbulent flow. With lower air velocities and air dilution, higher concentration of contamination can be established in these turbulent areas. A velocity in the region of 0.3 to 0.5 m/s (60 to 100 ft/min) is necessary so that disrupted unidirectional flow can be quickly reinstated and the contamination in turbulent areas round obstructions adequately diluted.

I have studied the effect of velocity in a variable-velocity unidirectional flow room that was operational. The velocity could be varied from 0.1 to 0.6 m/s (20 to 120 ft/min). The results showed that a velocity of 0.3 m/s (60 ft/min) was required to give stable unidirectional flow and low particle and bacterial concentrations. Increasing the air velocity, in stages up to 0.6 m/s (120 ft/min) gave lower airborne counts, but this worked on the "law of diminished returns". The information obtained can be interpreted as suggesting that a velocity of 0.3 m/s (60 ft/min) gives the best returns for effort, but if a cleanroom has a high density of machinery, or personnel, a higher velocity would give lower airborne contamination.

Unidirectional airflow is correctly defined in terms of air velocity as the higher the velocity the cleaner the room. Air changes per hour are related to the volume of the room, e.g. ceiling height and are therefore incorrect units of measurement.

The air supplied to unidirectional flow rooms is many times greater (10 s or 100 s of times) than that supplied to a turbulently ventilated room. These cleanrooms are therefore very much more expensive to build and run. Unidirectional flow rooms are of two general types, namely horizontal or vertical flow. In the horizontal system, the airflow is from wall to wall and in the vertical system it is from ceiling to floor.

## Vertical flow unidirectional cleanrooms

**A vertical flow unidirectional cleanroom** (垂直单向流洁净室) is shown in Fig. 12.6. This shows the air flowing through the complete area of a floor. However, unidirectional flow rooms are also designed so that air is returned through extract grilles distributed around the wall at floor level. This type is illustrated in Fig. 12.7. This design can only be used in rooms that are not too wide, and 6 meters (20 ft) has been suggested as a maximum width.

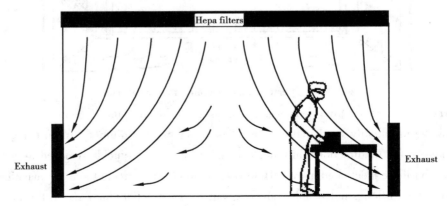

Fig. 12.7 Vertical unidirectional cleanroom with exhausts in the wall

Care must be taken with this design. The route that the supply air takes to get to the air exhausts is the reason for the problem. Airflow of the type shown in Fig. 12.7 gives poor unidirectional flow in the centre of the room and a flow of air that is not vertical in the rest. Personnel may therefore contaminate the product if they are positioned between the air supply and the product.

## Horizontal flow unidirectional cleanrooms

Fig. 12.8 shows a typical design of a **horizontal flow unidirectional cleanroom**（水平单向流洁净室）. In this design, the air is supplied through a wall of high efficiency filters and flows across the room and exits on the other side. The air is then returned to a ventilation plant and back through the air filters. The area of a wall in most rooms is usually much smaller than the ceiling, and hence a crossflow room will cost less in capital and running costs than a downflow one. The horizontal flow type of cleanroom is not as popular as the vertical type. The reason for this is illustrated in Fig. 12.9, which shows the problem that can occur with a contamination source.

Any contamination generated close to the filters in a vertical flow will be swept across the room and could contaminate work that is progressing downwind. Generally speaking, a vertical flow of air gives better contamination control (as shown in Fig. 12.9) because the dispersed contamination is less likely to reach the product.

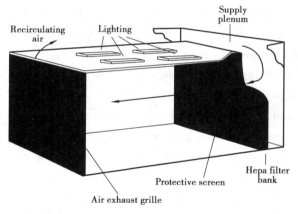

**Fig. 12.8　Horizontal flow unidirectional cleanrooms**

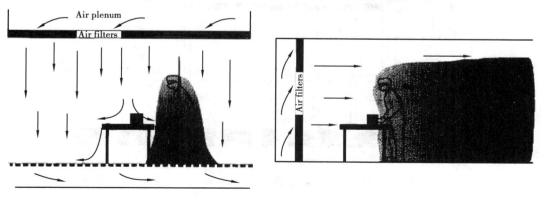

**Fig. 12.9　Dispersion in both a downflow and crossflow cleanroom**

If the crossflow cleanroom can be arranged so that the most critical operations are close to the supply filters and the dirtier ones at the exhaust end, then this type of room can be successful. The following works well:

- A faulty component, requiring repair, enters the end of the room away from the filters;
- The component is dismantled in stages as work progresses towards the filters;
- The most susceptible-to-contamination repair is carried out next to the supply filters;
- The component is reassembled and then packaged on its way back up the other side of the cleanroom;
- The repaired component exits out of the room, on the opposite side from the entering components.

A crossflow type of cleanroom can also be successful if the machine or process is placed close to the filter bank and no-one passes between the filter bank and machine when production is going on.

## Unidirectional flow rooms used in semiconductor manufacturing

Unidirectional flow cleanrooms are much used in semiconductor fabrication where the very best cleanroom conditions are required.

The design of semiconductor cleanrooms has evolved over several years, but a design that has been popular for a number of years is shown in Fig. 12.10. The air in Fig. 12.10 flows in a unidirectional way from a ceiling of high efficiency filters and down through the floor of the cleanroom. As the manufacturing of semiconductors is sensitive to vibration and anti-vibration measures are incorporated. Some designs return the air from just below the floor, while other designs (similar to the type shown in Fig. 12.10) use a large basement, that basement being used for services. Other designs have both the sub floor and a basement below. The design shown in Fig. 12.10 is often called a "ballroom" type because there is one large cleanroom. Typically, it is over 1,000 m$^2$ in floor area and some rooms are large enough to hold two football fields. It is expensive to run but adaptable.

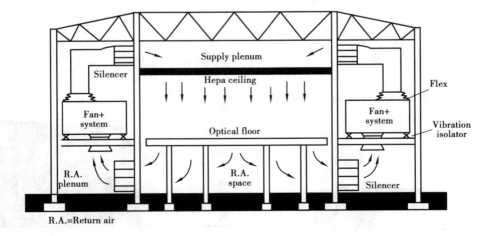

Fig. 12.10 Vertical flow cleanroom often used in semiconductor manufacturing

## 2. Clean air devices

Clean air devices are used in cleanrooms to provide a higher quality of air to critical areas where products or processes are open to contamination. Types of enhanced clean air devices that are available are: unidirectional flow benches, unidirectional flow work stations, isolators and **minienvironments**(微环境). The use of clean air devices in a turbulently ventilated room is popular, and is the most common configuration of cleanroom. By this method, the best air conditions are provided only where they are needed and considerable cost savings are made compared to providing a full unidirectional cleanroom.

### Unidirectional air devices

Fig. 12.11 is a drawing of a horizontal flow cabinet; this is one of the simplest and most effective methods of controlling contamination.

The operator sits at the bench and works on materials, or at a process, placed on the bench top. The operator's contamination is thus kept downwind of the critical process.

Also available are a variety of styles of unidirectional flow systems that may vary in size to encompass any size of production machinery.

### Isolators and minienvironments

A reduction in capital and running costs of a cleanroom is always sought, especially if this can be accompanied by an increase in product yield brought about by enhanced contamination control. There has therefore been much use of what have been variously called "isolators", "barrier technology" and "minienvironments". Minienvironments is the term commonly used in the semiconductor industry; the other two words are used mainly in the pharmaceutical industry.

Semiconductor applications. A minienvironment uses a physical barrier (usually plastic sheet or glass) to isolate the critical manufacturing area and provides it with the very best quality of air. The rest of the room can then be provided with lower quantities of air.

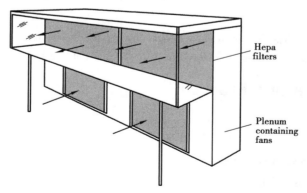

**Fig. 12.11 Horizontal flow cabinets**

Fig. 12.12 is a diagram of the air supply design without a minienvironment. In this design, large quantities of a unidirectional flow of air are provided to give the best conditions (shown as white in Fig. 12.12 and designated ISO Class 3) in those parts of the room where the operators move silicon wafers from machine to machine. Lesser quantities of air are provided for the service chases where the bulkhead-fitted machines are serviced (shaded area and designated ISO Class 6).

Fig. 12.13 is a diagram of an air supply design that uses minienvironments. The minienvironment (shown as white and designated as ISO Class 3) then provides the highest quality of environment. A lesser quality of environment (ISO Class 6 or poorer) is provided in both the production areas and the service chase. The total air supply volume is much less with this minienvironment design. It is seen in Fig. 12.13 that the air velocity in the minienvironment is shown as 0.4 m/s (80 ft/min). This is the speed associated with low particle counts in occupied cleanrooms. However, there are no personnel within the minienvironment causing turbulence and if the machinery is not giving off disruptive thermal up-currents then a lower air velocity may suffice.

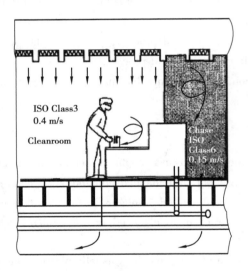
Fig. 12.12  Design of unidirectional system with service chase

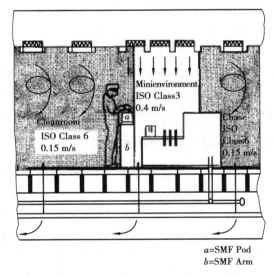
Fig. 12.13  Semiconductor fabrication room with a SMIF isolation system

**Source from:**

W. Whyte. Cleanroom Technology-Fundamentals of Design, Testing and Operation. chapter 6 Design of Unidirectional Cleanrooms and Clean Air Devices. England: John Wiley & Sons Ltd, 2001.

## Part 6    Exercise and Practice

### 1. *Discussions*

(1) Why do we need to design a cleanroom?

(2) How do we classify the cleanroom?

### 2. *Translation*

(1) Western countries have begun the era of air pollution as they used the coal instead of wood

in the fourteenth century, which belongs to the soot type and the first stage in air pollution.

(2) In the field of air cleaning, the particle concentration is typically very low compared with industrial dust separation applications, and the particle size is small. Hence the reliability of the performance of filters must be assured.

(3) Air filter is the main equipment in the field of air cleaning technology, and it is an indispensable equipment to create the clean air environment. So it is necessary to know the characteristic of air filters and its design principle so as to use it correctly and effectively.

(4) Only when indoor airborne particles move towards the nearby of the precision product and then deposit onto the sensitive area, damage may be caused for the product. So it is important to understand the mechanism of particle movement and deposition for the control of environment.

(5) Due to the limit of the means to sample the dust and the limit of the depth of understanding of air cleaning technology in early times, either the particle counting concentration or the particle weight concentration has been used to describe the air cleanliness.

## 3. Writing

Please explain the difference between unidirectional flow and nonunidirectional flow in 200 ~ 300 words.

# Lesson 13

# Refrigeration System

## Part 1　Preliminary

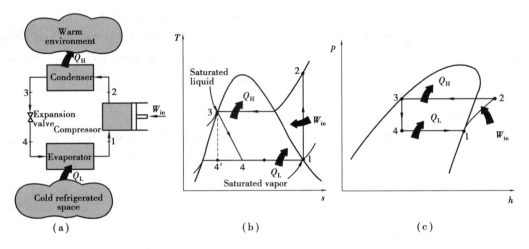

Fig. 13.1　Vapor-compression Refrigeration cycle

Vapor-compression refrigeration or vapor-compression refrigeration system (VCRS), in which the refrigerant undergoes phase changes, is one of the many refrigeration cycles and is the most widely used method for air-conditioning of buildings and automobiles. It is also used in domestic and commercial refrigerators, large-scale warehouses for chilled or frozen storage of foods and meats, refrigerated trucks and railroad cars, and a host of other commercial and industrial services. Refrigeration may be defined as lowering the temperature of an enclosed space by removing heat from that space and transferring it elsewhere. A device that performs this function may also be called an air conditioner, refrigerator, air source heat pump, geothermal heat pump or chiller (heat pump).

An absorption refrigerator is a refrigerator that uses a heat source (e. g. , solar energy, a fossil-fueled flame, waste heat from factories, or district heating systems) to provide the energy needed to drive the cooling process. The principle can also be used to air-condition buildings using the waste heat from a gas turbine or water heater. Using waste heat from a gas turbine makes the turbine very efficient because it first produces electricity, then hot water, and finally, air-conditioning (called cogeneration/trigeneration).

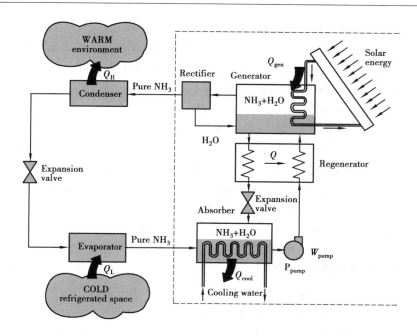

Fig. 13.2 **An absorption refrigeration cycle**

## Part 2  Text

蒸汽压缩式制冷的基本原理是利用液体在低压环境下蒸发吸收汽化潜热而实现的。本章节以逆卡诺循环为切入点,以蒸汽压缩式制冷装置的构成为基础,深入地阐述了蒸汽压缩式制冷的基本原理和工艺流程,分析了单级蒸汽压缩式制冷的热力循环、工质在 $T$-$s$ 图及 $p$-$h$ 图上的热力状态过程,以及制冷循环的热力计算过程。在此基础上,深入浅出地给出了多级压缩制冷的原理及方法,使读者对蒸汽压缩式制冷循环有概括性的认知。

Refrigeration was used by ancient civilizations when it was naturally available. The Roman rulers had slaves transport ice and snow from the high mountains to be used to preserve foods and to provide cool beverages in hot weathers. Today the refrigeration industry is a vast and essential part of any technological society. It is convenient to classify the applications of refrigeration into the following categories: domestic, commercial, industrial, and air conditioning.

## 1. *Refrigeration cycle*

Cooling means the removal of heat. In HVAC, a cooling process is usually identified as one which lowers the temperature or humidity (or both) of the ambient air. The effective temperature includes not only the temperature and humidity of the ambient air but also radiant effects and air movement. Some adiabatic cooling processes, i.e., evaporative cooling, do not actually remove any heat, but create a sensation of cooling by lowering the sensible temperature of the air.

A refrigeration cycle is a means of transferring heat from some place where it is not wanted (heat source) to another place where it can be used or disposed of (heat sink). The necessary components are (1) two or more **heat exchangers** ① (one each at source and sink), (2) a refrigerant, (3) a conduit for conveying the refrigerant, (4) mechanical and/or heat energy to move the refrigerant through the system, and (5) devices to control the rate of flow, to control temperature and pressure gradients, and to prevent damage to the system.

There are several basic refrigeration cycles. The two most common—two-phase (vapor compression) mechanical, and single-and double-effect absorption—are discussed below. Steam-jet refrigeration has historical importance but is not used in modern practice. Non-condensing (one-phase) mechanical cycles are used primarily in aircraft where light weight and simplicity are important. Thermoelectric refrigeration utilizes thermocouples working in reverse: When an electric current is impressed on a thermocouple, a cooling effect is obtained. These are small systems for specialized applications and are comparatively expensive to install and operate.

## 2. *Compression refrigeration cycle*

The most common cooling source in HVAC is mechanical two-phase vapor compression refrigeration. In this cycle (Fig. 13.3), a compressor is used to raise the pressure of a refrigerant gas. Work energy ($Q_W$) is required, usually provided by an electric motor or steam turbine or fuel-fired engine. The compression process raises the temperature of the gas. The high-pressure gas flows through piping to a condenser where heat is removed by transfer to a heat sink, usually water or air.

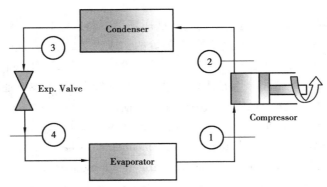

**Fig. 13.3 Basic Vapor-compression refrigeration cycle**

The refrigerant is selected with properties which allow it to condense (liquefy) at the temperature and pressure in the condenser. The high pressure liquid is passed through a pressure-reducing device to the evaporator. At the lower pressure, the liquid tends to evaporate, removing the heat of vaporization ($Q_C$) from its surroundings (the evaporator-technically, the heat source). The cold, low-pressure vapor is then returned to the compressor to be recycled. Note that the heat removed in the condenser is equal to the sum $Q_C + Q_W$. One index of refrigeration cycle effectiveness is its coefficient of performance (*COP*):

---

① 见 Part 3 Extension.

$$COP = Q_C/Q_W \qquad (13.1)$$

The refrigeration cycle can also be shown on a graph of the properties of a specific refrigerant. The graph in Fig. 13.1(c) is a pressure enthalpy or *p-h* diagram with the basic coordinates of pressure and enthalpy. The four stages of the cycle include compression (with a rise in temperature and enthalpy due to work done), condensing (cooling and liquefying at constant pressure), expansion (at constant enthalpy) and vaporization at constant pressure. The use of the *p-h* diagram allows the selection of the most effective refrigerant for the pressures and temperatures appropriate to the process. To minimize the work energy required, the temperature difference between the heat source and heat sink should be minimized.

## 3. *The basic vapor compression cycle*

As the refrigerant passes through the evaporator, heat transfer from the refrigerated space results in the vaporization of the refrigerant. Considering a control volume enclosing the refrigerant side of the evaporator (See Fig. 13.3), the rate of heat transfer per unit mass of refrigerant flow in the evaporator is as follows:

$$q_e = \frac{Q_e}{m} = h_1 - h_4 \qquad (13.2)$$

where $m$ is the mass flow rate of the refrigerant and $Q_e$ is the refrigeration capacity. The refrigeration capacity is usually expressed in kW or in tons of refrigeration.

Next consider the compressor.

$$w_i = \frac{W_i}{m} = h_2 - h_1 \qquad (13.3)$$

Where $W_i/m$ is the work done per unit mass of refrigerant.

Next, the refrigerant passes through the condenser, where the refrigerant condenses and there is heat transfer from the refrigerant to the cooler surroundings. For a control volume enclosing the refrigerant side of the condenser, the rate of heat transfer from the refrigerant per unit mass of refrigerant is:

$$q_c = \frac{Q_c}{m} = h_2 - h_3 \qquad (13.4)$$

Finally, the refrigerant at state 3 enters the **expansion valve**[②] and expands to the evaporator pressure. This process is usually modeled as a throttling process in which there is no heat transfer, i.e., for which

$$h_4 = h_3 \qquad (13.5)$$

$$COP = \frac{q_e}{w_i} = \frac{Q_e/m}{W_i/m} = \frac{h_1-h_4}{h_2-h_1} \qquad (13.6)$$

A theoretical single-stage cycle using R-134a as the refrigerant operates with a condensing temperature of 30 ℃ and evaporating temperature of -20 ℃. The system produces 50 kW of refrigera-

---

② 见 Part 3 Extension.

tion. Determine the thermodynamic property values at the four main state points of the cycle, the coefficient of performance of the cycle, the cycle refrigerating efficiency, and the rate of refrigerant flow. Fig. 13.4 shows a schematic p-h diagram for the problem with numerical property data. Saturated vapor and saturated liquid properties for states 1 and 3 are obtained from the saturation table for R-134a.

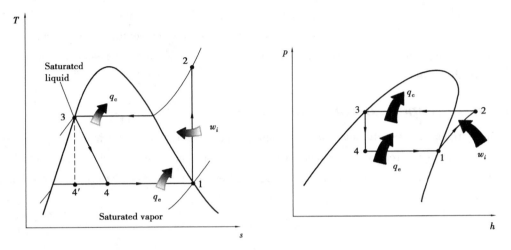

Fig. 13.4  **T-s and p-h diagram for basic vapor-compression refrigeration cycle**

Properties for superheated vapor at state 2 are obtained by linear interpolation of the superheat tables for R-134a. Specific volume and specific entropy values for state 4 are obtained by determining the quality of the liquid-vapor mixture from the enthalpy. The property data can be found in Table 13.1 based on the p-h diagram (Fig. 13.5).

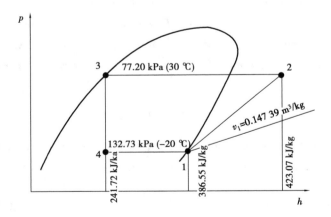

Fig. 13.5  **Schematic p-h diagram for the example**

$$x_4 = \frac{h_4 - h_f}{h_g - h_f} = \frac{241.72 - 173.64}{386.55 - 173.64} = 0.319\,8$$

$$v_4 = v_f + x_4(v_g - v_f) = 0.000\,736\,2 + 0.319\,8(0.147\,39 - 0.000\,736\,2) = 0.047\,64 \text{ m}^3/\text{kg}$$

$$s_4 = s_f + x_4(s_g - s_f) = 0.900\,2 + 0.319\,8(1.741\,3 - 0.900\,2) = 1.169\,18 \text{ kJ}/(\text{kg} \cdot \text{K})$$

Specific refrigerating effect: $q_e = h_1 - h_4 = 386.55 - 241.71 = 144.84$ kJ/kg

Specific heat transfer in condenser: $q_c = h_2 - h_3 = 423.07 - 241.72 = 181.35$ kJ/kg

Specific work of isentropic compression: $w_i = h_2 - h_1 = 423.07 - 386.55 = 36.52$ kJ/kg

Volumetric refrigerating effect: $q_v = (h_1 - h_3)/v_1 = \dfrac{144.84}{0.147\,39} = 982.70$ kJ/m$^3$

Volumetric work of isentropic compression: $w_v = (h_2 - h_1)/v_1 = \dfrac{36.52}{0.147\,39} = 247.78$ kJ/m$^3$

$COP = \dfrac{q_e}{w_i} = \dfrac{h_1 - h_4}{h_2 - h_1} = \dfrac{386.55 - 241.71}{423.07 - 386.55} = 3.97$

The mass flow of refrigerant is obtained from an energy balance on the evaporator. Thus,

$m(h_1 - h_4) = Q_e = 50$ kW

so, $m = \dfrac{Q_e}{h_1 - h_4} = \dfrac{50}{386.55 - 241.72} = 0.345$ kg/s

**Table 13.1 Thermodynamic property data for the example**

| state | $t(℃)$ | $p(kPa)$ | $v(m^3/kg)$ | $h(kJ/kg)$ | $s[kJ/(kg \cdot K)]$ |
|---|---|---|---|---|---|
| 1 | -20.0 | 132.73 | 0.147 39 | 386.55 | 1.741 3 |
| 2 | 37.8 | 77.20 | 0.027 98 | 423.07 | 1.741 3 |
| 3 | 30.0 | 77.20 | 0.000 842 | 241.72 | 1.143 5 |
| 4 | -20.0 | 132.73 | 0.047 636 | 241.72 | 1.169 18 |

The saturation temperatures of the single-stage cycle have a strong influence on the magnitude of the coefficient of performance. This influence may be readily appreciated by an area analysis on a temperature-entropy (T-s) diagram. The area under a reversible process line on a T-s diagram is directly proportional to the thermal energy added or removed from the working fluid. This observation follows directly from the definition of entropy.

## 4. Multi-stage vapor compression systems

The reasons for using a multistage vapor compression system instead of a single-stage system are as follows:

• The compression ratio of each stage in a multistage system is smaller than that in a single stage one, so compressor efficiency is increased.

• Liquid refrigerant enters the evaporator at a lower enthalpy and increases the refrigeration effect. Discharge gas from the high-stage compressor can have a lower temperature than the one-stage system at the same pressure difference between condensing and evaporating pressure. Multi-stage system can accommodate the variation of refrigeration load.

## 5. Two-stage expansion system with a flash cooler

A two-stage system is a refrigeration system working with a two-stage compression and mostly also with a two-stage expansion. A schematic system layout and the corresponding process in a log $p$-$h$ diagram are shown in Fig. 13.6.

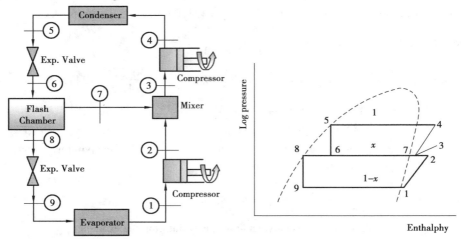

Fig. 13.6  Two stage expansion system with a flash cooler

## 6. Performance characteristics

In evaluating contributions of the compressor to thermodynamic systems, it is necessary to consider properties of the refrigerants at the inlet and outlet of the compressor, with the change in state between these points being (1) reversible and adiabatic (isentropic) for the ideal compressor; or (2) adiabatic and irreversible (with an increase in entropy in the fluid passing through the compressor) with the variation from the ideal compressor described by the adiabatic compressor efficiency.

An important thermodynamic consideration for the positive displacement compressor is the effect of the clearance volume, i. e. , the volume the refrigerant occupies within the compressor that is not displaced by the moving member. For the piston compressor consider the clearance volume between piston and cylinder head when the piston is in a top, center position. After the cylinder discharges the compressed gas, the clearance gas re-expands to a larger volume as the pressure falls to the inlet pressure. Consequently, the compressor discharges a refrigerant mass less than the mass that would occupy the volume swept by the piston, measured at the inlet pressure and temperature. This effect is quantitatively expressed by the volumetric efficiency

$$\eta_v = \frac{m_a}{m_t}$$

where,

$m_a$—actual mass of new gas entering the compressor per stroke;

$m_t$—theoretical mass of gas represented by the displacement volume and determined at the pressure and temperature at the compressor inlet.

Volumetric efficiency measures the effectiveness of the compressor's piston displacement (size) in moving the refrigerant vapor through the cycle. Since refrigerants differ greatly in their specific volumes $v_1$, choice of refrigerant can affect the mass flow delivered by compressor displacement.

One of the design parameters of a multistage compressor is selection of the interstate pressure at which the refrigerant temperature is reduced by an intercooler. At **optimum interstage pressure**[③], total work is minimum. For two- stage compression of an ideal gas ($pv = RT$), this occurs at the geometric mean of the suction and discharge pressures and results in equal work for the stages. The application of multistage compressors to refrigeration systems, however, differs from gas compressors since cooling at the interstage pressure is usually accomplished by refrigerant diverted from some other part of the cycle.

**Keywords**: COP, Refrigeration, Expansion, Refrigerant, Heat Calculation

**Source from**:
[1] 张寅平,潘毅群,王馨. 建筑环境与设备工程专业. Lesson 1 Thermodynamics and Refrigeration Cycle, Lesson 8: Cenreal Cooling and Heating. 北京:中国建筑工业出版社,2005.
[2] 赵三元,闫岫峰. 暖通与燃气. Lesson 4 The Ideal Basic Compression Refrigeration Cycle. 北京:中国建筑工业出版社,2002.
[3] Roger W. Haines, Lewis Wilson. HVAC Systems Design Handbook. Chapter 9 Cooling. Fourth Edition. McGraw-Hill Professional, April 1, 2003.

# Part 3  Extension

1. heat exchanger(换热器)

指的是冷凝器和蒸发器,冷凝器在高温侧,蒸发器位于低温侧,实际循环过程还有其他换热,如热回收器、中间冷却器等。

2. expansion valve(节流阀)

设置节流阀的制冷循环节流过程是一个等焓过程,此过程为不可逆过程;若设置膨胀机以替代膨胀阀则可回收膨胀功,则3-4过程即为等熵过程,为可逆过程。

3. optimum interstage pressure(最优中间压力)

中间压力对双级压缩制冷循环的经济性,如压缩机的容量、结构、功率和安全运行等都有直接影响。因此,合理地确定中间压力,是双级压缩制冷循环中的一个重要问题。根据不同情况,通常有两种确定方法:(1)按制冷系数最大为原则确定中间压力,这样得到的中间压力称为最佳中间压力。由于制冷循环形式不同,制冷系数的表达式也是不一样的,所以很难用一个统一的表达式进行计算。(2)按高、低压级压缩机的压缩比相等为原则,求得中间压力。计算式为 $p_M = \sqrt{p_1 \cdot p_2}$,其中 $p_M$ 为中间压力,$p_1$、$p_2$ 分别为高压压缩机和低压压缩机的压力。这种方法确定的中间压力对压缩机工作容积的利用程度较高,具有实用价值。

---

③ 见 Part 3 Extension.

## Part 4　Words and Expressions

| | |
|---|---|
| refrigeration | n. 制冷；冷藏；[热]冷却 |
| humidity | n. [气象]湿度；湿气 |
| adiabatic | adj. [物]绝热的；隔热的 |
| refrigerant | n. 制冷剂 |
| compressor | n. 压缩机 |
| absorption | n. 吸收 |
| condenser | n. 冷凝器 |
| evaporate | vi. 蒸发，挥发 |
| evaporator | n. 蒸发器 |
| expansion | n. 膨胀 |
| vaporization | n. 蒸发 |
| temperature difference | 温差 |
| capacity | n. 能力；容量 |
| expansion valve | 膨胀阀 |
| throttling process | 节流过程 |
| coefficient of performance (COP) | 性能系数 |
| schematic | adj. 图解的；概要的 |
| diagram | n. 图表；图解 |
| saturation | n. 饱和；色饱和度 |
| interpolation | n. 插入；插值 |
| isentropic | adj. [热]等熵的 |
| proportional | adj. 比例的，成比例的 |
| multistage | adj. 多级的 |
| variation | n. 变化 |
| piston | n. 活塞 |
| parameter | n. 参数；系数；参量 |
| minimum | n. 最小值；最低限度；最小化；最小量 |

## Part 5　Reading

### Refrigeration Equipments

#### 1. *Compressors*

In the refrigeration cycle, a compressor is a pump, providing the work energy to move the refrigerant from the low-pressure region to the high-pressure region through the system. Compressors come in two general types: positive displacement and centrifugal. Positive displacement compressors include reciprocating, rotary, scroll, and helical rotary (screw) types.

## Reciprocating compressors

Reciprocating compressors are usually the single-acting piston type. Figure 13.7 shows one possible arrangement. The volume swept by the piston is the displacement. The remaining volume under the cylinder head is the clearance. The theoretical volumetric efficiency is a function of the ratio of these two volumes together with the compression ratio of the system. Higher compressions ratios result in lower volumetric efficiencies. The actual volumetric efficiency will be somewhat less due to pressure drops across valves and other inefficiencies. The capacity of the compressor is a function of the volumetric efficiency, the properties of the refrigerant, and the operating pressures.

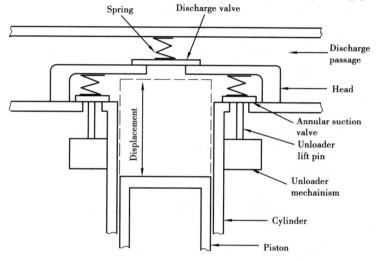

**Fig. 13.7 Reciprocating compressor**

A compressor is always designed for a specific refrigerant at some narrowly defined range of operating pressure. A compressor may have from 1 to 12 cylinders. Some older machines had as many as 24 cylinders. Most compressors for comfort cooling are direct-driven by electric motors. Historically, compressors were belt driven, often by steam engines and at slow speeds.

Sizes range up to as high as 200 tons or more in one compressor, although units over 100 tons are rare. If larger capacities are needed, two or more compressors are used in parallel. Small compressors (to 5 or 7.5 tons) are capacity-controlled by cycling the unit on and off. Larger units usually have unloaders on all but one or two cylinders. The unloader is activated electrically or pneumatically to lift the suction valve off its seat so that no compression takes place. Unloaders may be activated in stages so that two or more steps of capacity control may be obtained. Hot gas bypass is sometimes used with reciprocating compressors to maintain stable operation at reduced loads. Hot gas bypass does incur a power cost penalty.

Reciprocating compressors require from slightly less than 1 hp to as much as 1.5 hp/ton of actual refrigeration capacity at the maximum design temperatures and pressures typical of comfort-cooling processes. The horsepower per ton increases as the suction pressure and the temperature decrease. Therefore, it is more energy-efficient to operate at the highest suction pressure compatible with the needs of the application.

Compressors are lubricated by force-feed pumps or, in small units, by splash distribution of oil from the sump. Lubricating oils are selected to be miscible with the refrigerant and are carried throughout the piping system, which must be designed to ensure return of the oil to the compressor.

**Helical rotary compressors**

Helical rotary or screw compressors are made in single-screw and twin-screw types. The single-screw compressor consists of a helical main rotor with two star wheels. The enclosure of the main rotor has two slots through which the star wheel teeth pass; these teeth, together with the rotor and its enclosure, provide the boundaries of the compression chambers. The twin-screw compressor has two meshing helical gears and works much as a gear pump, with the helical shape forcing the gas to move in a direction parallel to the rotor shaft. These machines typically are direct-driven at 3,600 r/min and are usually oil-flooded for lubrication and to seal leakage paths. Capacity control is obtained by means of a sliding or rotating slotted valve.

**Centrifugal compressors**

Centrifugal compressors belong to the family of turbomachines, which includes fans and centrifugal pumps. Pressures and flows result from rotational forces. In HVAC work, these compressors are used primarily in package chillers where the compressors provide large capacities. Typical driven speed is 3,600 r/min or more, using electric motor engines or steam or gas turbines. Standard centrifugal chillers range in capacity from 100 to 2,000 tons, although some special units have been built with capacities as great as 8,500 tons. Capacity control is obtained by varying the driven speed or by means of inlet vanes, similar to those used on centrifugal fans. Noncondensing air-cycle systems, such as used on commercial aircraft, use high-speed gas-turbine drives at up to 90,000 r/min.

## 2. Chillers

The term chiller is used in connection with a complete chiller package—which includes the compressor, condenser, evaporator, internal piping, and controls—or for a liquid chiller (evaporator) only, where the water or brine is cooled. Liquid chillers come in two general types: flooded and direct expansion. There are several different configurations: shell-and-tube, double-tube, shell-and-coil, Baudelot (plate-type), and tank with raceway. For HVAC applications, the shell-and-tube configuration is most common.

**Flooded chillers**

A typical flooded shell-and-tube liquid chiller is shown in Fig. 13.8. Refrigerant flow to the shell is controlled by a high- or low-side float valve or by a restrictor. The water flow rate through the tubes is defined by the manufacturer but it generally ranges from 6 to 12 ft/s. Tubes may be plain (bare) or have a finned surface. The two-pass arrangement shown is most common, although one to four passes are available. The chiller must be arranged with removable water boxes so that the tubes may be cleaned at regular intervals, because even a small amount of fouling can cause a significant decrease in the heat exchange capacity. The condenser water tubes are especially subject to fouling with an open cooling tower. Piping must be arranged to allow easy removal of the water boxes.

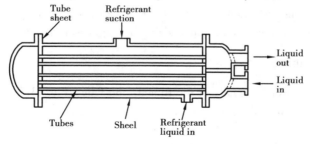

Fig. 13.8  **Flooded liquid chiller**

## Package chillers

A complete package chiller will include compressor, condenser, evaporator (chiller), internal piping, and operating and capacity controls. Controls should be in a panel and include all internal wiring with a terminal strip for external wiring connections. In small packages-up to about 250 tons-motor starters are usually included. In larger chillers, unit-mounted starters are an option. Some units with air-cooled condensers are designed for outdoor mounting; freeze prevention procedures must be followed. Units with water-cooled condensers require an external source of condensing water.

Chillers with reciprocating compressors are found mostly in the 5 to 200-tons range. Although larger units are made, economics usually favor centrifugal compressor or screw chillers, in sizes of 150 tons or more. Screw compressor systems are made in a wide range of sizes, by a growing number of manufacturers. With larger chillers or with high-voltage motors (2,300 V, 4,160 V), motor starters are usually separate from the centrifugal or screw chiller mounting frame and require field wiring of power and control circuits. Centrifugal compressor packages may be turbine-driven, occasionally engine-driven, but most often are driven by electric motors. The typical system is direct-driven at 3,600 r/min. Wye-delta motors are often used for reduced-voltage starting. Electronic "soft-start" devices are now available. Variable speed controllers are inherently soft starting. In larger units of 1,000 tons or more, it is not unusual to use high-voltage motors; the lower current requirements allow smaller wire sizes and across-the-line starting. An unusual drive system evolved on one of the 8,500-ton chillers at a major international airport. The utility plant manager replaced an original steam-turbine driver with a 5,000-hp 4,160-V variable-speed, variable-frequency electric drive. The chiller capacity was reduced to 5,500 tons, more in line with the actual load.

## 3. *Condensers*

The purpose of the condenser in a two-phase refrigeration cycle is to cool and condense the hot refrigerant gas leaving the compressor discharge. It is, then, a heat exchanger, of the shell-and-tube, tube-and-fin, or evaporative type. The heat sink is air, water, or a process liquid. Typically, small package systems use air-cooled condensers. Large built-up systems use water-cooled or evaporative condensers, although large air-cooled condensers are sometimes employed. The contrasting criteria here are the lower first cost of air-cooled condensers compared with the improved efficiencies obtained with water-cooled or evaporative condensers. The improvement in efficiency comes about because of the lower condensing temperatures achieved with water-cooled or evaporative condensers. The condenser capacity should match as closely as possible the capacity of the compressor in the system, although oversizing is preferable to undersizing if compressor efficiency is to be maximized.

## Air-cooled condensers

An air-cooled condenser is usually of the finned-tube type (Fig. 13.9), with the refrigerant in the tubes and air forced over the outside of the tubes and fins by a fan.

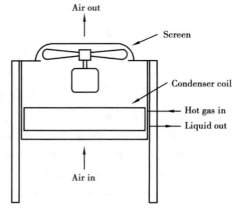

Fig. 13.9  Air-cooled condenser

Capacity control, if used, is accomplished by cycling the fan, using a multispeed fan, or modulating airflow by means of dampers. Refrigerant flow velocities must be designed to prevent oil traps in the tubes. Capacities are based on square feet of coil face area, fan airflow rate, desired condensing temperature, and design ambient dry-bulb (db) temperature. Note that at reduced ambient temperatures; performance of air-cooled condensers improves; sometimes approaching the seasonal performance of water-cooled systems.

## Water-cooled condensers

Water-cooled condensers are typically of the shell-and-tube type, with the water in the tubes and the refrigerant in the shell (Fig. 13.10). Chiller capacity control is not normally related to

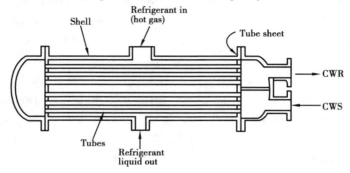

Fig. 13.10  Shell-and-tube, water-cooled condenser

condenser water temperature. However, the water temperature may vary if it is supplied from a cooling tower, and most chiller manufacturers prefer that condensing water not be taken below 65 to 70 °F because the oil may get held up in the condenser. Most code authorities do not allow direct use (and waste) of domestic water for condensing purposes.

**Source from:**

[1] 张寅平,潘毅群,王馨. 建筑环境与设备工程专业. Lesson 8 Central Cooling and Heating, Lesson 9 Decentralized Cooling and Heating. 北京:中国建筑工业出版社,2005.

[2] 赵三元,闫岫峰. 暖通与燃气. Lesson 4 The Ideal Basic Compression Refrigeration Cycle. 北京:中国建筑工业出版社,2002.

[3] Roger W. Haines, Lewis Wilson. HVAC Systems Design Handbook, Chapter 9 Cooling. Fourth Edition. McGraw-Hill Professional, 2003.

## Part 6  Exercises and Practices

### 1. *Discussion*

(1) Are there any difference and relationship between the compressible and absorption refrigeration systems?

(2) Please talk about how the evaporating and condensing temperatures affect the cycle quantities with your parterres.

(3) Are there any approaches to improve the refrigeration coefficient for both the compressible and absorption refrigeration systems? If yes, list your approaches as many as possible.

### 2. *Translation*

(1) A refrigeration cycle is a means of transferring heat from some place where it is not wanted (heat source) to another place where it can be used or disposed of (heat sink).

(2) The refrigerant is selected with properties which allow it to condense at the temperature and pressure in the condenser. The high pressure liquid is passed through a pressure-reducing device to the evaporator. At the lower pressure, the liquid tends to evaporate, removing the heat of vaporization from its surroundings.

(3) The saturation temperatures of the single-stage cycle have a strong influence on the magnitude of the coefficient of performance. This influence may be readily appreciated by an area analysis on a temperature-entropy ($T$-$s$) diagram. The area under a reversible process line on a $T$-$s$ diagram is directly proportional to the thermal energy added or removed from the working fluid. This observation follows directly from the definition of entropy.

(4) For the piston compressor consider the clearance volume between piston and cylinder head when the piston is in a top, center position. After the cylinder discharges the compressed gas, the clearance gas re-expands to a larger volume as the pressure falls to the inlet pressure.

### 3. *Writing*

Please write the processes and components of the compressible refrigeration cycle in about 200 words.

# Lesson 14
# Heat Pump

## Part 1  Preliminary

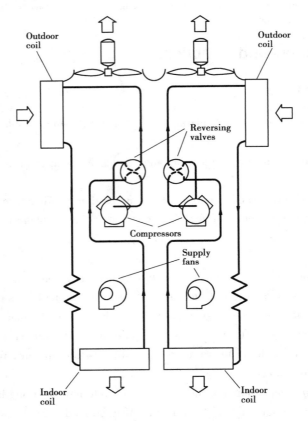

**Fig. 14.1  A typical rooftop heat pump**

In an air-source heat pump system, outdoor air acts as a heat source from which heat is extracted during heating, and as a heat sink to which heat is rejected during cooling. Since air is readily available everywhere, air-source heat pumps are the most widely used heat pumps in residential and many commercial buildings. The cooling capacity of most air-source heat pumps is between 1 and 30 tons (3.5 and 105 kW).

Most air-source heat pumps consist of single or multiple compressors, indoor coils through which air is conditioned, outdoor coils where heat is extracted from or rejected to the outdoor air, capillary tubes, reversing valves that change the heating operation to a cooling operation and vice versa, an accumulator to store liquid refrigerant, and other accessories. *

## Part 2　Text

热泵是一种利用高位能使热量从低位热源流向高位热源的节能装置。顾名思义，热泵也就是像泵那样，可以把不能直接利用的低位热能（如空气、土壤、水中所含的热能、太阳能、工业废热等）转换为可以利用的高位热能，从而达到节约部分高位能（如煤、燃气、油、电能等）的目的。热泵的种类很多，分类方法各不相同，可按热源种类、热泵驱动方式、用途、热泵工作原理、热泵工艺类型等方面来分类。在美国，对供建筑空调与供热用的热泵，按热源种类和热媒种类来划分。这种分类方法在我国也普遍采用。

### 1. Introduction

A heat pump extracts heat from a source and transfers it to a sink at a higher temperature. According to this definition, all pieces of refrigeration equipment, including air conditioners and chillers with refrigeration cycles, are heat pumps. In engineering, however, the term heat pump is generally reserved for equipment that heats for beneficial purposes, rather than that which removes heat for cooling only. Dual-mode heat pumps alternately provide heating or cooling. Heat reclaim heat pumps provide heating only, or simultaneous heating and cooling. An applied heat pump requires competent field engineering for the specific application, in contrast to the use of a manufacturer-designed unitary product. Applied heat pumps include built-up heat pumps (field-or custom-assembled from components) and industrial process heat pumps. Most modern heat pumps use a vapor compression (modified Rankine) cycle or absorption cycle. Although most heat pump compressors are powered by electric motors, limited use is also made of engine and turbine drives. Applied heat pumps are most commonly used for heating and cooling buildings, but they are gaining popularity for efficient domestic and service water heating, pool heating, and industrial process heating.

Applied heat pumps with capacities from 1.75 kW to 44 MW operate in many facilities. Some machines are capable of output water temperatures up to 105 ℃ and steam pressures up to 400 kPa (gage).

Compressors in large systems vary from one or more reciprocating or screw types to staged centrifugal types. A single or central system is often used, but in some instances, multiple heat pump systems are used to facilitate zoning. Heat sources include the ground, well water, surface water, gray water, solar energy, the air, and internal building heat. Compression can be single-stage or multistage. Frequently, heating and cooling are supplied simultaneously to separate zones.

Decentralized systems with water loop heat pumps are common, using multiple water-source heat pumps connected to a common circulating water loop. They can also include ground coupling, heat rejectors (cooling towers and dry coolers), supplementary heaters (boilers and steam heat exchangers), loop reclaim heat pumps, solar collection devices, and thermal storage. The initial cost

---

\* 来源于 Shan K. Wang. Handbook of Air Conditioning and Refrigeration-Chapter 12. New York: McGraw-Hill,2000.

is relatively low, and building reconfiguration and individual space temperature control are easy.

Community and district heating and cooling systems can be based on both centralized and distributed heat pump systems.

## 2. Heat pump cycles

Several types of applied heat pumps (both open-and closed-cycle) are available; some reverse their cycles to deliver both heating and cooling in HVAC systems, and others are for heating only in HVAC and industrial process applications. The following are the four basic types of heat pump cycles:

• Closed vapor compression cycle (Fig. 14.2). This is the most common type in both HVAC and industrial processes. It uses a conventional, separate refrigeration cycle that may be singlestage, compound, multistage, or cascade.

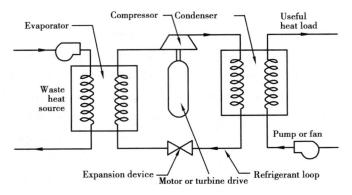

**Fig. 14.2 Closed vapor compression cycle**

• Mechanical vapor recompression (MVR) cycle with heat exchanger (Fig. 14.3). Process vapor is compressed to a temperature and pressure sufficient for reuse directly in a process. Energy consumption is minimal, because temperatures are optimum for the process. Typical applications include evaporators (concentrators) and distillation columns.

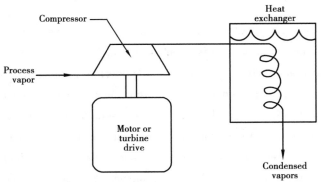

**Fig. 14.3 Mechanical vapor recompression cycle with heat exchanger**

• Open vapor recompression cycle (Fig. 14.4). A typical application is in an industrial plant with a series of steam pressure levels and an excess of steam at a lower-than-desired pressure. Heat is pumped to a higher pressure by compressing the lower-pressure steam.

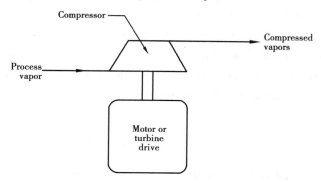

Fig. 14.4  **Open vapor recompression cycle**

• Heat-driven Rankine cycle (Fig. 14.5). This cycle is useful where large quantities of heat are wasted and energy costs are high. The heat pump portion of the cycle may be either open or closed, but the Rankine cycle is usually closed.

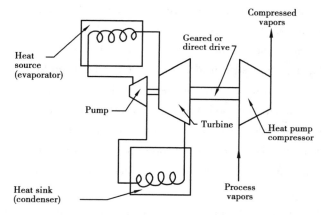

Fig. 14.5  **Heat-driven rankine cycle**

## 3. *Heat sources and sinks*

Selecting a heat source and sink for an application is primarily influenced by geographic location, climate, initial cost, availability, and type of structure.

### Air

Outdoor air is a universal heat source and sink medium for heat pumps and is widely used in residential and light commercial systems. Extended-surface, forced-convection heat transfer coils transfer heat between the air and refrigerant. Typically, the surface area of outdoor coils is 50 to 100% larger than that of indoor coils. The volume of outdoor air handled is also greater than the volume of indoor air handled by about the same percentage. During heating, the temperature of the evaporating refrigerant is generally 6 to 11 K less than the outdoor air temperature.

When selecting or designing an **air-source heat pump**[①], two factors in particular must be consideed: (1) the local outdoor air temperature and (2) frost formation.

As the outdoor temperature decreases, the heating capacity of an air-source heat pump decreases. This makes equipment selection for a given outdoor heating design temperature more critical for an air-source heat pump than for a fuel-fired system. Equipment must be sized for as low a balance point as is practical for heating without having excessive and unnecessary cooling capacity during the summer.

## Water

Water can be a satisfactory heat source. City water is seldom used because of cost and municipal restrictions. Groundwater (well water) is particularly attractive as a heat source because of its relatively high and nearly constant temperature. Water temperature depends on source depth and climate, but, in the United States, generally ranges from 5 ℃ in northern areas to 20 ℃ in southern areas. Frequently, sufficient water is available from wells (water can be reinjected into the aquifer). This use is nonconsumptive and, with proper design, only the water temperature changes. Water quality should be analyzed, and the possibility of scale formation and corrosion should be considered. In some instances, it may be necessary to separate the well fluid from the equipment with an additional heat exchanger. Special consideration must also be given to filtering and settling ponds for specific fluids. Other considerations are the costs of drilling, piping, pumping, and a means for disposal of used water. Information on well water availability, temperature, and chemical and physical analysis is available from U.S. Geological Survey offices in many major cities.

Heat exchangers may also be submerged in open ponds, lakes, or streams. When surface or stream water is used as a source, the temperature drop across the evaporator in winter may need to be limited to prevent freeze-up.

In industrial applications, waste process water (e.g., spent warm water in laundries, plant effluent, warm condenser water) may be a heat source for heat pump operation.

Sewage, which often has temperatures higher than that of surface or groundwater, may be an acceptable heat source. Secondary effluent (treated sewage) is usually preferred, but untreated sewage may be used successfully with proper heat exchanger design.

Use of water during cooling follows the conventional practice for water-cooled condensers.

Water-to-refrigerant heat exchangers are generally direct-expansion or flooded water coolers, usually shell-and-coil or shell-and-tube. Brazed-plate heat exchangers may also be used. In large applied heat pumps, the water is usually reversed instead of the refrigerant.

## Ground

The ground is used extensively as a heat source and sink, with heat transfer through buried coils. Soil composition, which varies widely from wet clay to sandy soil, has a predominant effect on

---

① 见 Part 3 Extension.

thermal properties and expected overall performance. The heat transfer process in soil depends on transient heat flow. Thermal diffusivity is a dominant factor and is difficult to determine without local soil data. Thermal diffusivity is the ratio of thermal conductivity to the product of density and specific heat. The soil's moisture content influences its thermal conductivity.

There are three primary types of ground-source heat pumps: (1) groundwater, which is discussed in the previous section; (2) direct-expansion, in which the ground-to-refrigerant heat exchanger is buried underground; and (3) ground-coupled (also called closed-loop ground-source), in which a secondary loop with a brine connects the ground-to-water and water-to-refrigerant heat exchangers (see Tab. 14.1).

Ground loops can be placed either horizontally or vertically. A horizontal system consists of single or multiple serpentine heat exchanger pipes buried 1 to 2 m apart in a horizontal plane at a depth 1 to 2 m below grade. Pipes may be buried deeper, but excavation costs and temperature must be considered. Horizontal systems can also use coiled loops referred to as slinky coils. A vertical system uses a concentric tube or U-tube heat exchanger.

**Solar Energy**

Solar energy may be used either as the primary heat source or in combination with other sources. Air, surface water, shallow groundwater, and shallow ground-source systems all use solar energy indirectly. The principal advantage of using solar energy directly is that, when available, it provides heat at a higher temperature than the indirect sources, increasing the heating coefficient of performance. Compared to solar heating without a heat pump, the collector efficiency and capacity are increased because a lower collector temperature is required.

Research and development of solar-source heat pumps has been concerned with two basic types of systems: direct and indirect. The direct system places refrigerant evaporator tubes in a solar collector, usually a flat-plate type. Research shows that a collector without glass cover plates can also extract heat from the outdoor air. The same surface may then serve as a condenser using outdoor air as a heat sink for cooling.

An indirect system circulates either water or air through the solar collector. When air is used, the collector may be controlled in such a way that (1) the collector can serve as an outdoor air preheater; (2) the outdoor air loop can be closed so that all source heat is derived from the sun, or (3) the collector can be disconnected from the outdoor air serving as the source or sink.

## 4. *Types of heat pumps*

Heat pumps are classified by (1) heat source and sink; (2) heating and cooling distribution fluid; (3) thermodynamic cycle; (4) building structure; (5) size and configuration; and (6) limitation of the source and sink. Tab. 14.1 shows the more common types of closed vapor-compression cycle heat pumps for heating and cooling service.

**Air-to-Air Heat Pumps**. This type of heat pump is the most common and is particularly suitable for factory-built unitary heat pumps. It is widely used in residential and commercial applications. The first diagram in Tab. 14.1 is a typical refrigeration circuit.

In air-to-air heat pump systems, air circuits can be interchanged by motor-driven or manually operated dampers to obtain either heated or cooled air for the conditioned space. In this system, one heat exchanger coil is always the evaporator, and the other is always the condenser. Conditioned air passes over the evaporator during the cooling cycle, and outdoor air passes over the condenser. Damper positioning causes the change from cooling to heating.

**Water-to-Air Heat Pumps.** These heat pumps rely on water as the heat source and sink, and use air to transmit heat to or from the conditioned space. (See the second diagram in Tab. 14.1.) They include the following:

- **Groundwater heat pumps**, which use groundwater from wells as a heat source and/or sink. They can either circulate source water directly to the heat pump or use an intermediate fluid in a closed loop, similar to the ground-coupled heat pump.

- **Surface water heat pumps**[②], which use surface water from a lake, pond, or stream as a heat source or sink. As with ground-coupled and groundwater heat pumps, these systems can either circulate source water directly to the heat pump or use an intermediate fluid in a closed loop.

- Internal-source heat pumps, which use the high internal cooling load generated in modern buildings either directly or with storage. These include water-loop heat pumps.

- **Solar-assisted heat pumps**, which rely on low-temperature solar energy as the heat source. Solar heat pumps may resemble water-to-air, or other types, depending on the form of solar heat collector and the type of heating and cooling distribution system.

- **Wastewater-source heat pumps**, which use sanitary waste heat or laundry waste heat as a heat source. Waste fluid can be introduced directly into the heat pump evaporator after waste filtration, or it can be taken from a storage tank, depending on the application. An intermediate loop may also be used for heat transfer between the evaporator and the waste heat source.

**Water-to-Water Heat Pumps.** These heat pumps use water as the heat source and sink for cooling and heating. Heating/cooling changeover can be done in the refrigerant circuit, but it is often more convenient to perform the switching in the water circuits, as shown in the third diagram of Tab. 14.1. Although the diagram shows direct admittance of the water source to the evaporator, in some cases, it may be necessary to apply the water source indirectly through a heat exchanger (or double-wall evaporator) to avoid contaminating the closed chilled-water system, which is normally treated. Another method uses a closed-circuit condenser water system.

**Ground-coupled heat pumps**[③]. These use the ground as a heat source and sink. A heat pump may have a refrigerant-to-water heat exchanger or may be direct-expansion (DX). Both types are shown in Tab. 14.1. In systems with refrigerant-to-water heat exchangers, a water or antifreeze solution is pumped through horizontal, vertical, or coiled pipes embedded in the ground. Direct-expansion ground-coupled heat pumps use refrigerant in direct-expansion, flooded, or recirculation evaporator circuits for the ground pipe coils.

---

②③ 见 Part 3 Extension.

Tab. 14.1 **Heat pump types**

| Heat source and sink | Distribution fluid | Thermal cycle |
|---|---|---|
| Air | Air | Refrigerant changeover |
| Water | Air | Refrigerant changeover |
| Water | Water | Water changeover |
| Ground-coupled (or Closed-loop ground-source) | Air | Refrigerant changeover |
| Ground-source, direct-expansion | Air | Refrigerant changeover |

Soil type, moisture content, composition, density, and uniformity close to the surrounding field areas affect the success of this method of heat exchange. With some piping materials, the material of construction for the pipe and the corrosiveness of the local soil and underground water may affect the heat transfer and service life. In a variation of this cycle, all or part of the heat from the evaporator plus the heat of compression are transferred to a water-cooled condenser. This condenser heat is then available for uses such as heating air or domestic hot water.

Additional heat pump types include the following:

Air-to-Water Heat Pumps Without Changeover. These are commonly called heat pump water heaters.

Refrigerant-to-Water Heat Pumps. These condense a refrigerant by the cascade principle. Cascading pumps the heat to a higher temperature, where it is rejected to water or another liquid. This type of heat pump can also serve as a condensing unit to cool almost any fluid or process. More than one heat source can be used to offset those times when insufficient heat is available from the primary source.

**Keywords**: Air-source heat pump, Groundwater heat pumps, Ground-coupled heat pumps

**Source from**:
ASHRAE. 2008 ASHRAE Handbook. HVAC Systems and Equipment-Chapter 8. Atlanta, GA: ASHRAE, 2008.

## Part 3  Extension

1. Air-source heat pump(空气源热泵)

热泵的功能是把热从低位势(低温端)抽升到高位势(高温端)排放。空气源热泵就是利用室外空气的能量通过机械做功,使能量从低位热源向高位热源转移的制冷/热装置。它以冷凝器放出的热量来供热,以蒸发器吸收的热量来供冷。

2. Surface water heat pumps(地表水源热泵)

地表水源热泵是以地表水体为低位热源,利用水源热泵机组为空调系统制备与提供冷/热水,再通过空调末端设备实现房间空气调节的系统。作为低位热源的地表水体,可以利用温度合适的地表水,含海水、湖水、江河水等。

3. Ground-Coupled Heat Pumps(土壤耦合式热泵)

土壤耦合式热泵系统一般由地埋管换热器、水源热泵机组和室内末端系统3部分组成。在夏季,地埋管内的传热介质(水或防冻液)通过水泵送入冷凝器,将热泵机组排放的热量带走并释放给地层(向大地排热,地层为蓄热);蒸发器中产生的冷水,通过循环水泵送至空调末端设备,对房间进行供冷。在冬季,热泵机组通过地下埋管吸收地层的热量(从土壤吸热,地层为蓄冷),冷凝器产生的热水,则通过循环水泵送至空调末端设备,对房间进行供暖。

## Part 4  Words and Expressions

| | |
|---|---|
| coil | n. 盘管 |
| compressor | n. 压缩机 |
| reversing valve | 换向阀 |
| supply fan | 送风机 |
| heat pump | 热泵 |
| air conditioner | 空调,空调器 |
| refrigeration cycles | 制冷循环 |
| dual-mode | 双模态;双重模式 |

续表

| | |
|---|---|
| heat reclaim | 热回收，热量回收 |
| vapor compression cycle | 蒸汽压缩循环 |
| absorption cycle | 吸收式循环 |
| turbine drive | 汽轮机驱动 |
| heat pump cycles | 热泵循环 |
| evaporator | n. 蒸发器 |
| condenser | n. 冷凝器 |
| heat source | 热源 |
| heat sink | 热汇 |
| air-source heat pump | 空气源热泵 |
| frost formation | 结霜 |
| corrosion | n. 腐蚀，侵蚀，锈蚀；受腐蚀的部位；衰败 |
| heat exchanger | 换热器 |
| filtering | n. 过滤，滤除，滤清<br>v. 透过（filter 的现在分词）；（光或声）渗入 |
| sewage | n. 污水；下水道；下水道里的）污物 |
| direct-expansion | 直接膨胀式 |
| soil | n. 土壤 |
| wet clay | 湿黏土 |
| sandy soil | 沙土；砂土 |
| thermal properties | 热性质 |
| transient heat flow | 瞬态热流 |
| thermal diffusivity | 热扩散系数 |
| solar energy | 太阳能 |
| collector efficiency | 集热器效率 |
| thermodynamic cycle | 热力循环 |
| air-to-air heat pumps | 空气-空气热泵 |
| water-to-air heat pumps | 水-空气热泵 |
| groundwater heat pumps | 地下水源热泵 |
| surface water heat pumps | 地表水源热泵 |
| solar-assisted heat pumps | 太阳能辅助热泵 |
| wastewater-source heat pumps | 污水源热泵 |
| water-to-water heat pumps | 水-水热泵 |
| ground-coupled heat pumps | 土壤耦合式热泵 |

# Part 5  Reading

## Groundwater Heat Pump System

Groundwater heat pump (GWHP) systems use well water as a heat source during heating and as a heat sink during cooling. When the groundwater is more than 30 ft (9 m) deep, its year-round temperature is fairy constant. Groundwater heat pump systems are usually open-loop systems. They are mainly used in low-rise residences in northern climates such as New York or North Dakota. Sometimes they are used for low-rise commercial buildings where groundwater is readily available and local codes permit such usage.

## 1. *Groundwater systems*

For commercial buildings, the design engineer must perform a survey and study the site and surroundings to define the available groundwater sources. The design engineer should be fully aware of the legalities of water rights.

A test well should be drilled to ensure the availability of groundwater. If water is corrosive, a plate-and-frame heat exchanger may be installed to separate the groundwater and the water entering the water coil in the water-source heat pump (WSHP).

Usually, two wells are drilled. One is the supply well, from which groundwater is extracted by submersible pump impellers and supplied to the WSHPs. The other well is a recharge or injection well. Groundwater discharged from the WSHP is recharged to this well. The recharged well should be at least 100 ft (30 m) away from the supply well. Using a recharge well provides for resupply to the groundwater and prevents the collapse of the building foundation near the supply well due to subsidence. If the quality of groundwater meets the requirement and if local codes permit, groundwater discharged from the WSHP can be used as the service water or can be drained to the nearby river, lake, or canal. The groundwater intake water screen of the supply well may be located in several levels where water can be extracted. For a small supply well for residences, the pump motor is directly connected to the submersible pump underneath the impellers, whereas for large wells, the motor is usually located at the top of the supply well. Information regarding groundwater use regulations and guidelines for well separation and for the construction of supply and recharge wells are included in Donald's (1985) Water Source Heat Pump Handbook, published by the National Water Well Association (NWWA).

If the temperature of the groundwater is below 50 °F (10 °C), direct cooling of the air in the WSHP should be considered. If the groundwater temperature exceeds 55 °F (12.8 °C) and is lower than 70 °F (21.1 °C), precooling of recirculating air or makeup air may be economical.

Because the groundwater system is an open-loop system, it is important to minimize the vertical head to save pump power. Air-conditioning and Refrigeration Institute (ARI) Standard 325-85 recommends that the groundwater pump power not exceed 60 W/gpm (950 W s /L). If the pump efficiency is 0.7, the allowable head for the well pump, including static head, head loss across the water coil or heat exchanger, valves, and piping work losses, should preferably not exceed 220 ft WC (66 m WC). If a recharge well is used and the discharge pipe is submerged under the water table level in the recharge well, as shown in Fig. 4.6, the groundwater system is most probably a closed-loop system, depending on whether the underground water passage between the supply and recharge well is connected or broken.

## 2. *Groundwater heat pump systems for residences*

A typical groundwater heat pump system for a residence is shown in Fig. 14.6. Such a heat pump system usually has a rated heating capacity from 24,000 to 60,000 Btu/h (7,030 to 17,580 W). Groundwater is extracted from a supply well by means of a submersible well pump and is forced through a precooler or direct cooler. Then groundwater enters the water coil in the water-source heat pump. After that, groundwater is discharged to a recharge well. If the recharge pipe is submerged underneath the water table, as described in the previous section, such a groundwater system is most probably a closed-loop system. In the recharge well, the water level is raised in order to overcome the head loss required to force the groundwater discharging from the perforated pipe wall through the water passage underground. The vertical head required is the difference between the water levels in the supply and recharge wells, as shown in Fig. 14.6.

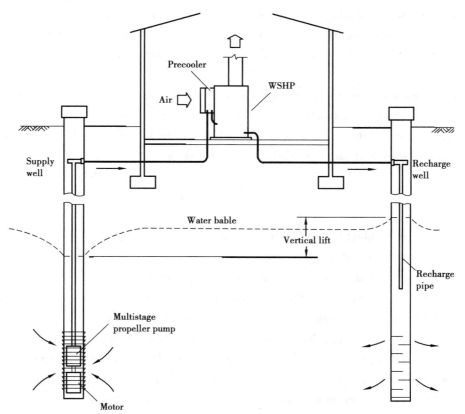

**Fig. 14.6 A typical residential ground water heat pump system**

Operating parameters and characteristics for a groundwater heat pump system are as follows:

- The groundwater flow rate for a water-source heat pump should vary from 2 to 3 gpm per 12,000 Btu/h (0.036 to 0.054 L/s per kW) heating capacity. The well pump must be properly sized. A greater flow rate and an oversized well pump are not economical.
- The pressure drop of the groundwater system should be minimized. The pressure drop per 100 ft (30 m) of pipe should be less than 5 ft/100 ft (5 m/100 m) of length. Unnecessary valves should not be installed. Gate valves or ball valves should be used instead of globe valves to reduce pressure loss. A water tank is not necessary.
- Direct cooling or precooling of recirculating air by means of groundwater increases significantly the EER of the GWHP system.
- Water containing excessive concentrations of minerals causes deposits on heat pump water coils that reduce the heat pump performance.
- In locations where the number of annual heating degree-days HDD65 exceeds 7,000, more than 80 percent of the operating hours of the GWHP systems are for space heating.
- An electric heater may be used for supplementary heating in cold climates or in other locations where it is necessary. When the heating capacity of GWHP is equal to or even greater than the heating load operation of the electric heater must be avoided.
- Water-source heat pumps should be properly sized. They should not be operated for short-cycle durations (cycles less than 5 min) in order to prevent cycling losses and excessive wear and tear on the refrigeration system components. Cycle durations between 10 and 30 min are considered appropriate.
- Extraction and discharge of groundwater must comply with local codes and regulations.
- The temperature of groundwater tends to increase with its use.

A parameter called the seasonal performance factor (SPF) is often used to assess the performance of a groundwater heat pump system. SPF is defined as

$$SPF = \frac{Q_{\text{sup}}}{Q_{\text{cons}}} \tag{14.1}$$

Where,

$Q_{\text{sup}}$—all heat supplied by GWHP during heating season, Btu (kJ),

$Q_{\text{cons}}$—energy consumed during heating season, Btu (kJ).

For groundwater heat systems with a cooling capacity $Q_{\text{rc}}$ < 135,000 Btu/h (40 kW), ASHRAE/IESNA Standard 90.1-1999 mandates the minimum efficiency requirements of 11.0 EER for water-source heat pumps using groundwater during cooling mode when the entering water is 70 °F (21.1 ℃). As of October 29, 2001, the minimum efficiency requirement is 16.2 EER when the entering water is 59 °F (15 ℃). During heating mode, the minimum efficiency requirement is 3.4 COP when the entering water temperature is 70 °F (21.1 ℃). As of October 29, 2001, the minimum efficiency requirement is 3.6 COP when the entering water temperature is 50 °F (10 ℃).

A groundwater heat pump system has a fairly constant COP even if the outdoor air temperature varies. According to Rackliffe and Schabel (1986), the SPF and EER for 15 single-family houses in New York State from 1982 to 1984 were as follows:

    System SPF (heating)                  1.9 to 3 average 2.3

    System average EER (cooling)       5.6 to 14 averages 9.2

In many locations, groundwater heat pump systems usually have a higher SPF and seasonal EER than air-source heat pumps. GWHP system capacity remains fairly constant at very low and very high outdoor temperatures.

The main disadvantage of a groundwater heat pump system is its higher initial cost. More maintenance is required for systems using water with high mineral content. If the water table is 200 ft (60 m) or more below ground level, the residential groundwater heat pump system is no longer energy-efficient compared to high-efficiency air-source heat pumps.

**Source from:**
Shan K. Wang. Handbook of Air Conditioning and Refrigeration-Chapter 12. New York: McGraw-Hill, 2000.

## Part 6    Exercises and Practices

### 1. *Discussion*

    (1) Why are the heat pumps so effective?

    (2) Discuss the common types of heat pump.

### 2. *Translation*

    (1) Applied heat pumps are most commonly used for heating and cooling buildings, but they are gaining popularity for efficient domestic and service water heating, pool heating, and industrial process heating.

    (2) Selecting a heat source and sink for an application is primarily influenced by geographic location, climate, initial cost, availability, and type of structure.

    (3) As the outdoor temperature decreases, the heating capacity of an air-source heat pump decreases. This makes equipment selection for a given outdoor heating design temperature more critical for an air-source heat pump than for a fuel-fired system.

    (4) Soil composition, which varies widely from wet clay to sandy soil, has a predominant effect on thermal properties and expected overall performance.

    (5) Surface water heat pumps, which use surface water from a lake, pond, or stream as a heat source or sink. As with ground-coupled and groundwater heat pumps, these systems can either circulate source water directly to the heat pump or use an intermediate fluid in a closed loop.

### 3. *Writing*

    Please describe the main advantages and disadvantages of a groundwater heat pump system in 200~300 words.

# Lesson 15
# Energy Resource

## Part 1  Preliminary

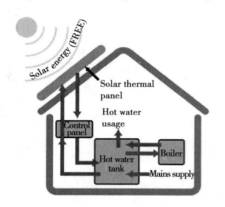

Fig. 15.1　**Solar thermal system**

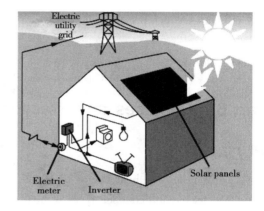

Fig. 15.2　**Solar electric system**

Solar energy can be used to generate electricity, provide hot water, and to heat, cold, and light buildings. Solar water heating systems for buildings have two main parts: a solar collector and a storage tank. The sun heats an absorber plate in the collector, which, in turn, heats the fluid running through tubes within the collector. Photovoltaic systems produce direct current (DC) power which fluctuates with the sunlight's intensity. For practical use this usually requires conversion to certain desired voltages or alternating current (AC), through the use of inverters. Many residential PV systems are connected to the grid wherever available.

## Part 2  Text

主动式的太阳能利用主要包括太阳能光热系统和太阳能光电系统。太阳能光热系统是指利用太阳能集热器、储水箱、循环泵以及控制装置等实现太阳能的收集、储存和热水利用。集热器主要有平板式、真空管式和聚光式。太阳能光热系统可以提供生活热水、冬季供暖和夏季热驱动制冷等。太阳能光伏发电可直接将太阳能转化成电能。此类系统中最常见的形式是和公共电网相连构成并网系统,还有一种非常经济的应用形式是为耗电量较少的公共设施比如标志、交通指示灯等提供电力。此外,太阳能光电系统可以和建筑材料相结合成为建筑结构的一部分。

Solar energy is the primary energy source that fuels the growth of the Earth's natural capital and drives wind and ocean currents that also can provide alternative energy sources. Since the beginning of time, solar energy has been successfully harnessed for human use. Early civilizations and some modern ones used solar energy for many purposes: food and clothes drying, heating water for baths, heating adobe and stone dwellings, etc. Solar energy is free and available to anyone who wishes to use it.

## 1. *Introduction*

Solar thermal heating for domestic hot-water and space heating has grown considerably over the years and is well established in several countries.

For many, a key impediment to increased solar use is economics. The cost of some solar technologies is perceived to be high compared to the fossil-based energy source it is offsetting. For example, while the simple payback of **solar PV systems**[①] tends to be rather long (although much improvement has occurred over the past couple of decades), the recent increase in the cost of energy and advances in solar energy justify a fresh look at the applications and the engineering behind those applications. In addition, public policy in many areas encourages solar and other renewable energy applications through tax incentives and encouraging or requiring repurchase of excess electrical energy generated by the utility provider.

The applicability and, consequently, the economics and public policy incentives available with different solar energy system types and applications depend greatly on the location, which determines the technical factors (such as solar resources available), as well as the nontechnical (such as the country's political situation).

## 2. *Solar thermal applications*

Solar energy thermal applications range from low-temperature applications (e.g., swimming pool heating or domestic water) to medium-to high-temperature applications (e.g., space heating, absorption cooling or steam production for electrical generation).

The most common solar energy thermal application is for domestic hot water (DHW) production. However, the same solar collectors can be used to deliver thermal energy for space heating. A typical installation for the combined production of DHW and space heating (i.e., solar combi systems) includes the solar collectors, the heat storage tank, and a boiler used as an auxiliary heater. Combi systems require a larger collector area than a DHW system to meet the higher loads. It is possible to use a heat storage tank and a DHW storage vessel, but it may also be suitable to combine them in a single storage tank (with a high vertical stratification) to meet the different operating temperatures for space heating and DHW. To assess and compare performances of different designs for solar combi systems, the International Energy Agency (IEA) launched Task 26 to address issues in this area. Standardized classification and evaluation processes and design tools were developed for these systems, along with proposals for the international standardization of combi system test procedures.

The main drawback of solar combi systems has been the fact that during summer, the available

---

① 见 Part 3 Extension.

high solar radiation and the heat produced from the solar collectors cannot be fully used, thus making the system financially less attractive and limiting its use to the low DHW summer demand. In addition, there are some technical problems related to stagnation (i.e. the condition when the medium in the solar collector loop vaporizes as a result of high solar radiation availability and low thermal demand). Since high building cooling loads generally coincide with high solar radiation, the readily available solar heat from the existing solar collectors can be exploited by a heat-driven cooling machine, thus extending the use of the solar field throughout the year (solar hot water [SHW] and space heating in winter and SHW and cooling in summer). Combining solar heating and cooling is usually referred to as a solar combi-plus system that can increase the total solar fraction.

The capital cost of solar thermal systems generally increases with higher working fluid temperatures. The higher the delivery temperature, the lower the efficiency, and the more solar collector area is generally required to deliver the same net energy. This is due to parasitic thermal losses that are inherent in solar collector design. Different solar collector types provide advantages and disadvantages, depending on the application, and there are significant cost differences among each solar collector type. Some common collector types are discussed here.

**Flat-Plate Solar Collectors**. These are best suited for processes requiring low-temperature working fluids (80 to 160 °F or 27 to 71 °C) and can deliver 80 °F (27 °C) fluid temperatures, even during overcast conditions. The term flat-plate collector generally refers to a hydronic coil-covered absorber housed in an insulated box with a single- or double-glass cover that allows solar energy to heat the absorber. Heat is removed by a fluid running through the hydronic coils. Its design makes it more susceptible to parasitic losses than an evacuated tube collector, but more efficient in solar energy capture, because flat-plate collectors convert both direct and indirect solar radiation into thermal energy. This makes flat-plate collectors the preferred choice for domestic hot-water and other low-temperature heating applications. Coupling a water-source heat pump with low fluid temperatures with solar collectors provides heating efficiencies that are higher than ground-source heat pump (GSHP) applications and standard natural gas furnaces.

**Evacuated-Tube Collectors**. These are a series of small absorbers consisting of small-diameter (approximately 3/8 in. or 10 mm) copper tubing encased in a clear, cylindrical, evacuated thermos bottle that minimizes parasitic losses even at elevated temperatures. Because of the relatively small absorber area, significantly more collector area is required than with the flat-plate or concentrating collector.

**Concentrating Collectors**. This refers to the use of a parabolic reflector that focuses the solar radiation falling within the reflector area onto a centrally located absorber. Concentrating collectors are best suited for processes requiring high-temperature working fluids (300 to 750 °F or 150 to 400 °C) and do not operate under overcast conditions. This type of collector converts only direct solar radiation, which varies dramatically with sky clearness and air quality (e.g., smog). Concentrating collectors rotate on one or two axes to track the sun and collect the available direct solar radiation.

The percentage of energy a solar system can provide is known as **the solar fraction**[②]. The F-Chart method developed by Sanford Klein at the University of Wisconsin provides an accurate assessment of the amount of energy a solar thermal system will provide. This modeling provides the designer the ability to vary system parameters (e. g., collector area, storage volume, operating temperature, and load) to optimize system design. The method was originally developed to aid engineers who did not have computer resources available to do the complex analyses required for solar thermal calculations. This method has since been adapted for automated computer analysis using modern tools.

The cost-effectiveness of thermal solar systems is also dependent on having a constant load for the energy the solar system provides. Since space heating requirements are generally seasonal in most climates, it is advisable that energy from the solar system have more uses than space heating alone. Domestic water heating is usually a much steadier year-round load, though typically not very substantial. Solar cooling can extend utilization of solar collectors during the summer. The fact that peak cooling demand in summer is associated with high solar radiation availability offers an excellent opportunity to exploit solar energy with heat-driven cooling machines.

Solar hot water can be used as the sole source of domestic hot water or to preheat incoming supply water. This can be as simple as having an uninsulated tank in a hot attic in a southern climate to having a batch water heater on the roof of a building. Pool heating is the reason for the vast majority of current sales of hot-water-based systems in most markets in the United States. Water is taken from the pool and pumped through an unglazed collector and then returned back into the pool. In many climates, most of the energy for pool heating can be offset with this technique. Solar energy is a major source for domestic hot-water heating in several countries (i. e., China, Israel, Australia, and Greece) and is currently widely used in states such as Hawaii and Florida. Although it is an economically viable option for practically all areas of the United States, a barrier to increased market acceptance is the current lack of trained installation contractors and limited information available in most areas.

Besides preheating water, solar energy can be used to preheat incoming air—before it is introduced to a building—using systems integrated with the overall building design or collectors. This is a simple technological approach that can be economically viable and is well suited for northern climates or areas with high building heating load. The DOE has published a bulletin on these transpired solar collectors.

Solar-assisted cooling systems employ solar thermal collectors connected to thermal-driven cooling devices. They consist of several main components, namely, the solar collectors, heat storage, heat distribution system, heat-driven cooling unit, optional cold storage, an air-conditioning system with appropriate cold distribution, and an auxiliary subsystem (which is integrated at different places in the overall system and is used as an auxiliary heater parallel to the collector or the collector/storage as an auxiliary cooling device, or both).

Due to their low first cost, common flat-plate solar collectors are used for reaching a driving

---

② 见 Part 3 Extension.

temperature of 140 °F to 194 °F (60 °F to 90 °C). With selective-surface, flat-plate solar collectors, the driving temperature can be up to 248 °F (120 °C); however, the collector efficiency will be quite low at this temperature level. Stationary flat-plate, evacuated tube solar collectors are typically used for 175 °F to 250 °F (79 °F to 121 °C) applications, and can reach higher temperatures but at a lower collector efficiency. Compound parabolic concentrators can reach 207 °F to 329 °F (97 °C to 165 °C).

Each technology has specific characteristics that match the building's HVAC design, loads, and local climatic conditions. A good design must first exploit all available solar radiation and then cover the remaining loads from conventional sources. Proper calculations for collector and storage size depend on the employed solar cooling technology. Hot-water storage may be integrated between the solar collectors and the heat-driven chiller to dampen the fluctuations in the return temperature of the hot water from the chiller. The storage size depends on the application: if cooling loads mainly occur during the day, then a smaller storage unit will be necessary than when the loads peak in the evening. Heating the hot-water storage by the backup heat source should be strictly avoided. The storage's only function is to store excess heat of the solar system and to make it available when sufficient solar heat is not available.

## 3. *Solar-electric systems*

The direct conversion from sunlight to electricity is accomplished with PV systems. These systems continue to drop in price, and many states (as well as the federal government) offer tax incentives for the installation of these systems. The most common application is a grid-tied system where electricity is directly fed into the grid.

Other applications of PV may be attractive and provide additional value beyond the cost of offsetting utility power. Whenever power lines must be extended, PV should be investigated. It is cost-effective to use stand-alone PV applications for signs, remote lights, and blinking traffic lights. Many times, if an existing small grid line is available and additional power is needed, it is less expensive to add the PV system rather than increase the power line. This may be true of remote guard shakes, restroom facilities, or other outbuildings. PV should also be considered as part of the uninterruptible power system for a building. Every building has battery backup for certain equipment. Batteries can be centralized and the PV fed directly into this system. The cost can be less than other on-site generation and fuel storage.

Technical and legal factors have generally been worked out that have hindered the use of grid-tied solar PV systems in the past. In temperate climates, grid-tied PV systems provide a good method to reduce peak summer electrical demand, since peak solar gains generally correspond to peak air-conditioning demand. This was a major factor that led to the passing of the California Million Solar Roofs initiative in early 2006. The German 100,000 rooftops-HTDP program, which began in early 1999, had a goal of 300 MW. It was successfully completed in 2003 with the parallel introduction of the German Renewable Energy Resources Act (EEG) that came into effect in 2000. The EEG and HTDP program secured commercially oriented PV investors (because they received a full payback of their investment).

The economic viability of PV systems is dependent on many factors. Improvements in technology have lowered the cost per peak watt from $5.14 and $3.08 in 1989 (in nominal dollars) to $3.49 for modules and $1.94 for cells in 2008, according to the DOE.

The module cost represents about 50% ~ 60% of the total installed cost of a solar energy system. Therefore, the solar module price is the key element in the total price of an installed solar system. All prices are exclusive of sales taxes, which, depending on the country or region, can add 8% ~ 20% to the prices, with the highest sales tax rates being in Europe. The typical PV cell efficiencies (i.e., the ratio of electrical energy produced by a solar cell to the incident solar irradiance) currently range from less than 5% for the first generation of thin-film cells to more than 24% for the most advanced crystalline-silicon cells in laboratory conditions.

The integration of PV systems with building materials is a new development that may, in the long run, help make PV systems viable in most areas. One of the key ways to incorporate solar energy technologies is to incorporate them directly into the architecture of the building. **Trombe walls**[③] can replace spandrel panels and other glass facades; PV panels can become overhangs for the building or parking shading structures. Using these elements as part of the building doubles the value of the systems. An example of these technologies is the concept called building-integrated PVs. These are being developed by the National Renewable Energy Laboratory (NREL), among others.

**Keywords**: Solar thermal energy system, solar electric system, integration of solar system with building

**Source from**:

ASHRAE. ASHRAE GreenGuide: The Design, Construction, and Operation of Sustainable Building. Atlanta, GA: ASHRAE, 2010.

# Part 3  Extension

1. Solar PV system(太阳能光电系统)

PV, Photovoltaic (solar cell),光伏发电。光电系统能将太阳光直接转化成电力。太阳能电池或光电电池是由能吸收太阳光的半导体材料组成的。太阳能激发半导体材料中的电子逃逸原子核,使之定向运动,产生电流。光电电池通常组合成模块,每个模块包含40个电池单元。大约10个模块安装在一个光电电池阵列上。光电电池阵列能用于为单座建筑提供电力,也能大规模地组成太阳能发电厂。

2. Solar fraction(太阳能保证率)

太阳能保证率是指来自太阳辐射的有效得热与供暖系统所需热负荷之比,即系统中由太阳能部分提供的热量除以系统总负荷。

---

③ 见 Part 3 Extension.

### 3. Trombe wall(特隆布墙)

特隆布墙也称为特隆布式集热墙,是20世纪60年代由法国工程师特隆布首次将这种集热理念发展为建筑构件的。典型的集热墙一般是由透明玻璃层、空气夹层和墙体几部分组成。针对不同需要,集热墙可分为多种类型,如开孔和不开孔(开孔形式又可分为内开孔和外开孔)。集热墙是利用阳光照射到外面有玻璃罩的深色蓄热墙体上,加热透明玻璃和厚墙外表面之间的夹层空气,通过热压作用使空气流入室内向室内供热,同时墙体本身直接通过热传导向室内放热并储存部分能量,夜间墙体储存的能量释放到室内。另一方面,集热墙也会通过玻璃层等将热量以传导、对流及辐射的方式传递到室外。

## Part 4  Words and Expressions

| | |
|---|---|
| harness | n. 背带;马具,挽具;系带 |
| | vt. 利用;给(马等)套轭具;控制 |
| dwelling | n. 居住;住处,处所;寓所 |
| | v. 居住,住(dwell 的现在分词) |
| domestic hot-water | 生活热水,卫生热水 |
| impediment | n. 障碍物;妨碍,阻止;口吃,结巴 |
| fossil-based energy source | 化石能源 |
| absorption cooling | 吸收式制冷 |
| stagnation | n. 滞止;淤塞,停滞;不景气 |
| coincide | vi. 相符;与……一致;想法、意见等相同 |
| heat-drivencooling | 热驱动制冷 |
| capital cost | 初投资 |
| parasitic | adj. 寄生的;寄生物的;由寄生虫引起的 |
| flat-Plate Solar Collectors | 平板式太阳能集热器 |
| overcast | adj. 天阴的,多云的 |
| | vt. 使沮丧;包缝;遮蔽 |
| | n. [气象学]阴,阴天;遮盖物(尤指云、雾等) |
| insulated box | 保温箱 |
| evacuated-Tube Collectors | 真空管式太阳能集热器 |
| encase | vi. 包装;围绕;把……装箱 |
| cylindrical | adj. 圆筒形;圆柱形的,圆筒状的,气缸的 |
| evacuated thermos bottle | 真空保温瓶 |
| concentrating collectors | 聚光式太阳能集热器 |
| parabolicreflector | 抛物面反射器 |
| axes | n. 轴;轴线 |
| attic | n. 阁楼;顶楼 |

续表

| | |
|---|---|
| viable | *adj.* 切实可行的；能养活的；有望实现的 |
| contractor | *n.* 承包商；承包人；收缩物 |
| solar-assisted cooling system | 太阳能辅助供冷系统 |
| auxiliary subsystem | 辅助子系统 |
| stationary | *adj.* 固定的；静止的，不变的；常备军的；定居的 |
| | *n.* 不动的人；驻军；固定物 |
| compound | *n.* 复合物；混合物；化合物 |
| | *vt.* 调和；使混合；使严重；调停 |
| grid-tied system | 并网系统 |
| utility power | 公用电力 |
| uninterruptible power system | 不间断电源系统 |
| hinder | *vt.&vi.* 阻碍，妨碍；成为阻碍 |
| | *adj.* 后面的，后方的 |
| thin-film cells | 薄膜电池 |
| crystalline-silicon cells | 晶体硅电池 |
| incorporate | *vt.* 包含；组成公司；使混合；使具体化 |
| | *vi.* 合并；包含；吸收；混合 |
| spandrel | *n.* 拱肩；拱脊 |

# Part 5　Reading

## Energy Resources

### 1. *Characteristics of energy and energy resource forms*

The HVAC & R industry deals with energy forms as they occur on or arrive at a building site. Generally, these forms are fossil fuels (natural gas, oil, and coal) and electricity. Solar and wind energy are also available at most sites, as is low-level geothermal energy (an energy source for heat pumps). Direct-use (high-temperature) geothermal energy is available at some locations.

**Forms of on-site energy**

Fossil fuels and electricity are commodities that are usually metered or measured for payment at the facility's location. Solar or wind energy is freely available but does incur cost for the means to use it. High-temperature geothermal energy, which is not universally available, may or may not be a sold commodity, depending on the particular locale and local regulations.

Some on-site energy forms require further processing or conversion into more suitable forms for the particular systems and equipment in a building or facility. For instance, natural gas or oil is burned in a boiler to produce steam or hot water, which is then distributed to various use points (e.g., heating coils in air-handling systems, unit heaters, convectors, fin-tube elements, steam-powered cooling units, humidifiers, kitchen equipment) throughout the building. Although the methods and efficiencies of these processes fall within the scope of the HVAC&R designer, how an energy source arrives at a given facility site is not under direct control. On-site energy choices, if available, may be controlled by the designer based in part on the present and projected future availability of the resources.

The basic energy source for heating may be natural gas, oil, coal, or electricity. Cooling may be produced by electricity, thermal energy, or natural gas. If electricity is generated on site, the generator may be driven by an engine or fuel cell that consumes fossil fuels or hydrogen on site, or by a turbine using steam or gas directly.

The term energy source refers to on-site energy in the form in which it arrives at or occurs on a site (e.g., electricity, gas, oil, coal). Energy resources refers to the raw energy that (1) is extracted from the earth (wellhead or mine-mouth), (2) is used to generate the energy source delivered to a building site (e.g., coal used to generate electricity), or (3) occurs naturally and is available at a site (solar, wind, or geothermal energy).

## Nonrenewable and renewable energy resources

From the standpoint of energy conservation, energy resources can be classified as either (1) nonrenewable resources, which have definite, although sometimes unknown, limitations; or (2) renewable resources, which have the potential to regenerate in a reasonable period. Resources used most in industrialized countries are nonrenewable.

Note that renewable does not mean an infinite supply. For instance, hydropower is limited by rainfall and appropriate sites, usable geothermal energy is available only in limited areas, and crops are limited by the available farm area and competing nonenergy land uses. Other forms of renewable energy also have supply limitations.

Nonrenewable resources of energy include:
- Coal;
- Crude oil;
- Natural gas;
- Uranium or plutonium (nuclear energy).

Renewable resources of energy include:
- Hydropower;
- Solar;
- Wind;
- Earth heat (geothermal);
- Biomass (wood, wood wastes, and municipal solid waste, landfill methane, etc.);
- Tidal power;

- Ocean thermal;
- Atmosphere or large body of water (as used by the heat pump);
- Crops (for alcohol production or as boiler fuel).

## Characteristics of fossil fuels and electricity

Most on-site energy for buildings in developed countries involves electricity and fossil fuels as primary on-site energy sources. Both fossil fuels and electricity can be described by their energy content (joules). This implies that energy forms are comparable and that an equivalence can be established. In reality, however, they are only comparable in energy terms when they are used to generate heat. Fossil fuels, for example, cannot directly drive motors or energize light bulbs. Conversely, electricity gives off heat as a byproduct regardless of whether it is used for running a motor or lighting a light bulb, and regardless of whether that heat is needed. Thus, electricity and fossil fuels have different characteristics, uses, and capabilities aside from any differences in their derivation.

Other differences between energy forms include methods of extraction, transformation, transportation, and delivery, and characteristics of the resource itself. Natural gas arrives at the site in virtually the same form in which it was extracted from the earth. Oil is processed (distilled) before arriving at the site; having been extracted as crude oil, it arrives at a given site as, for example, No. 2 oil or diesel fuel. Electricity is created (converted) from a different energy form, often a fossil fuel, which itself may first be converted to a thermal form. The total electricity conversion, or generation, process includes energy losses governed largely by the laws of thermodynamics.

Fuel cells, which are used only on a small scale, convert a fossil fuel to electricity by chemical means.

Fossil fuels undergo a conversion process by combustion (oxidation) and heat transfer to thermal energy in the form of steam or hot water. The conversion equipment is a boiler or a furnace in lieu of a generator, and conversion usually occurs on a project site rather than off-site. (District heating or cooling is an exception.) Inefficiencies of the fossil fuel conversion occur on site, whereas inefficiencies of most electricity generation occur off site, before the electricity arrives at the building site. (Cogeneration is an exception.)

Sustainability is an important consideration for energy use. The United Nations' Brundtland Report (UN 1987) stated that the development of the built environment is sustainable if it "meets the needs of the present without compromising the ability of future generations to meet their own needs".

## 2. *On-site energy/energy resource relationships*

An HVAC&R designer must select one or more forms of energy. Most often, these are fossil fuels and electricity, although installations are sometimes designed using a single energy source (e.g., only a fossil fuel or only electricity).

Solar energy normally impinges on the site (and on the facilities to be put there), so it affects the facility's energy consumption. The designer must account for this effect and may have to decide whether to make active use of solar energy. Other naturally occurring and distributed renewable forms such as wind power and earth heat (if available) might also be considered.

The designer should be aware of the relationship between on-site energy sources and raw energy resources, including how these resources are used and what they are used for. The relationship between energy sources and energy resources involves two parts: (1) quantifying the energy resource units expended and (2) considering the societal effect of depletion of one energy resource (caused by on-site energy use) with respect to others.

## Quantifiable relationships

As on-site energy sources are consumed, a corresponding amount of resources are consumed to produce that on-site energy. For instance, for every volume of No. 2 oil consumed by a boiler at a building site, some greater volume of crude oil is extracted from the earth. On leaving the well, the crude oil is transported and processed into its final form, perhaps stored, and then transported to the site where it will be used.

Even though natural gas often requires no significant processing, it is transported, often over long distances, to reach its final destination, which causes some energy loss. Electricity may have as its raw energy resource a fossil fuel, uranium, or an elevated body of water (hydroelectric generating plant).

Data are available to help determine the amount of resource use per delivered on-site energy source unit. In the United States, data are available from entities within the U. S. Department of Energy and from some agencies and associations.

A resource utilization factor (RUF) is the ratio of resources consumed to energy delivered (for each form of energy) to a building site. Specific RUFs may be determined for various energy sources normally consumed on site, including nonrenewable sources such as coal, gas, oil, and electricity, and renewable sources such as solar, geothermal, waste, and wood energy. With electricity, which may derive from several resources depending on the particular fuel mix of the generating stations in the region served, the over all RUF is the weighted combination of individual factors applicable to electricity and a particular energy resource. Grumman (1984) gives specific formulas for calculating RUFs.

There are great differences in the efficiency of equipment used in buildings. Although electricity incurs losses in its production, it is often much more efficient than direct fuel use at the building site, particularly for lighting or heat pump applications. Minimizing both energy cost and the amount of energy resources needed to accomplish a task effectively should be a major design goal, which requires consideration of both RUFs and end-use efficiency of building equipment.

Although a designer is usually not required to determine the amount of energy resources attributable to a given building or building site for its design or operation, this information may be helpful when assessing the long-range availability of energy for a building or the building's effect on energy resources. Fuel-quantity-to-energy resource ratios or factors are often used, which suggests that energy resources are of concern to the HVAC&R industry.

## Intangible relationships

Energy resources should not simply be converted into common energy units (e.g., gigajoule) because the commonality gives a misleading picture of the equivalence of these resources. Other differences and limitations of each of the resources defy easy quantification. For instance, electricity that arrives and is used on a site can be generated from coal, oil, natural gas, uranium, or hydro-power. The end result is the same: electricity at $x$ kV, $y$ Hz. However, the societal impact of a megajoule of electricity generated by hydro-power may not equal that of a megajoule generated by coal, uranium, domestic oil, or imported oil.

Intangible factors such as safety, environmental acceptability, availability, and national interest also are affected in different ways by the consumption of each resource. Heiman (1984) proposes a procedure for weighting the following intangible factors:

National/Global Considerations:
- Balance of trade;
- Environmental impacts;
- International policy;
- Employment;
- Minority employment;
- Availability;
- Alternative uses;
- National defense;
- Domestic policy;
- Effect on capital markets.

Local Considerations:
- Exterior environmental impact;
- Air;
- Solid waste;
- Water resources;
- Local employment;
- Local balance of trade;
- Use of distribution infrastructure;
- Local energy independence;
- Land use;
- Exterior safety.

Site Considerations:
- Reliability of supply;
- Indoor air quality;
- Aesthetics;
- Interior safety;
- Anticipated changes in energy resource prices.

## 3. Summary

In HVAC&R system design, the need to address immediate issues such as economics, performance, and space constraints often prevents designers from fully considering the energy resources affected. Today's energy resources are less certain because of issues such as availability, safety, national interest, environmental concerns, and the world political situation. As a result, the reliability, economics, and continuity of many common energy resources over the potential life of a building being designed are unclear. For this reason, the designer of building energy systems must consider the energy resources on which the long-term operation of the building will depend. If the continued viability of those resources is reason for concern, the design should provide for, account for, or address such an eventuality.

**Source from:**
ASHRAE. 2009 ASHRAE Handbook. Fundamentals-Chapter 34. Atlanta, GA: ASHRAE, 2009.

## Part 6  Exercises and Practices

### 1. Discussion

(1) How to apply renewable energy in buildings? And what are application barriers?

(2) Please describe the forms of solar thermal application.

### 2. Translation

(1) Solar energy thermal applications range from low-temperature applications (e.g., swimming pool heating or domestic water) to medium-to high-temperature applications (e.g., space heating, absorption cooling or steam production for electrical generation).

(2) It is possible touse a heat storage tank and a DHW storage vessel, but it may also be suitable to combine them in a single storage tank (with a high vertical stratification) to meet the different operating temperatures for space heating and DHW.

(3) The main drawback of solar combi systems has been the fact that during summer, the available high solar radiation and the heat produced from the solar collectors cannot be fully used, thus making the system financially less attractive and limiting its use to the low DHW summer demand.

(4) Each technology has specific characteristics that match the building's HVAC design, loads, and local climatic conditions.

(5) Other applications of PV may be attractive and provide additional value beyond the cost of offsetting utility power.

### 3. Writing

Please describe the characteristics of building energy consumptions in 200~300 words.

# Lesson 16

# Introduction of Fuel Gas

## Part 1  Preliminary

**Tab 16.1  The common fuel as composition**

| NO. | Sort of gases | Gas composition volume fraction $\Phi/\%$ | | | | | | | | |
|---|---|---|---|---|---|---|---|---|---|---|
| | | $CH_4$ | $C_3H_8$ | $C_4H_{10}$ | $C_mH_n$ | CO | $H_2$ | $CO_2$ | $O_2$ | $N_2$ |
| 1 | Natural gas | | | | | | | | | |
| | Pure natural gas | 98 | 0.3 | 0.3 | 0.4 | | | | | 1.0 |
| | Oil field gas | 81.7 | 6.2 | 4.86 | 4.94 | | | 0.3 | 0.2 | 1.8 |
| | Condensate field gas | 74.3 | 6.75 | 1.87 | 14.91 | | | 1.62 | | 0.55 |
| | Mine drainage gas | 52.4 | | | | | | 4.6 | 7.0 | 36.0 |
| 2 | Artificial gas | | | | | | | | | |
| 2-1 | Solid fuel destructive distillation gas | | | | | | | | | |
| | Coke oven gas | 27 | | | 2 | 6 | 56 | 3 | 1 | 5 |
| | Continuousvertical coking furnace gas | 18 | | | 1.7 | 17 | 56 | 5 | 0.3 | 2 |
| | Upright box furnace gas | 25 | | | | 9.5 | 55 | 6 | 0.5 | 4 |
| 2-2 | Solid fuel vaporizing gas | | | | | | | | | |
| | Pressure evaporation gas | 18 | | | 0.7 | 18 | 56 | 3 | 0.3 | 4 |
| | Water gas | 1.2 | | | | 34.4 | 52.0 | 8.2 | 0.2 | 4.0 |
| | Producer gas | 1.8 | | | 0.4 | 30.4 | 8.4 | 2.4 | 0.2 | 56.4 |
| 2-3 | Oil gas | | | | | | | | | |
| | Heavy oil thermalcracking gas | 28.5 | | | 32.17 | 2.68 | 31.51 | 2.13 | 0.62 | 2.39 |
| | Heavy oil catalytic pyrolysis gas | 16.6 | | | 5 | 17.2 | 16.5 | 7.0 | 1.0 | 6.7 |
| 2-4 | Blast-furnacegas | 0.3 | | | | 28.0 | 2.7 | 10.5 | | 58.5 |
| 3 | Liquefied petroleum gas (estimated value) | | 50 | 50 | | | | | | |
| 4 | Methane (biogas) | 60 | | | | a little | a little | 35 | a little | |

The fuel gas is mainly composed of a combustible part, which generally comprises hydrocarbons, hydrogen and carbon monoxide. And the non-combustible components are carbon dioxide, nitrogen and oxygen. There are many sorts of gas. And the natural gas and liquefied petroleum gas are the main source in the town, which are gradually substituting for the artificial gas. Biogas can be the fuel source for rural areas. Regardless of natural gas, liquefied petroleum gas or artificial gas, the composition and calorific value of the same kind of gas are different depending on the place of production. Table 16.1 shows the common composition of different gases.

## Part 2　Text

　　建设资源节约型与环境友好型社会的提出、国家对城市燃气领域的开放,以及管道建设的延伸,为中国城市燃气的发展提供了难得的机遇。我国有较为丰富的天然气资源,但存在地区分布不均衡的问题,陆上天然气资源主要分布在中西部和近海地区,约80%的陆上天然气分布在塔里木盆地、四川盆地、鄂尔多斯盆地、渤海湾盆地、准噶尔盆地和东南海域等地区。液化石油气作为管输天然气很好的补充气源,目前主要来自炼油厂。人工煤气是以煤或石油系产品为原料转化得到的可燃气体。生物气主要是各种有机物质在隔绝空气的条件下发酵产生的可燃气体。

　　This part will introduce some primary definitions and processes about the fuel gas, which is applicable to all natural gas systems and to liquefied petroleumgas (LPG) and LPG/air systems where appropriate. All pressures are gauge pressures unless otherwise stated.

## 1. *Definitions of the gas industry*

### Fuel gas

　　Any mixture of combustible gases.

### Gas supplier

　　In relation to gas means:
　　● a person who supplies gas to any premises through a primary meter;
　　● a person who provides a supply of gas to a consumer by means of the filling or re-filling of a storage container designed to be filled or re-filled with gas at the place where it is connectedfor use whether or not such container is or remains the property of the supplier;
　　● a person who provides gas in re-fillable cylinders for use by a consumer whether or not such cylinders are filled or re-filled directly by that person and whether or not such cylindersare or remain the property of that person, but a retailer shall not be deemed to be a supplierwhen he sells a brand of gas other than his own.

### Gas system

　　Gas system is comprised of a distribution main/service (pipe), emergency control valve, meter installation and installation pipework and any additional emergency control valve (AECV) to supply a consumer's appliance.

### Gas tight

　　Gas tight mean the property which is attributed to a component when it stops or contains gas flow under specified Conditions.

### Gas transporter (GT)

　　Companies which transport gas through its network on behalf of a gas shipper.

## Gas plant and equipment

All gas-containing components, including filters, heaters, slam-shut valves, regulators, relief valves, non-return valves (NRVs), meters, valves, pilots, burners, interconnecting pipework, instruments, impulse and control lines.

## Gas fitting

Gas pipework, valves (other than the emergency control valve, ECV), regulators, meters, fittings, apparatus and appliances designed for use by consumers of gas for heating, lighting, cooking or other purposes for which gas can be used, but it does not mean:
- any part of a distribution main or service (pipe);
- any part of a pipeline upstream of a distribution main or service (pipe);
- a gas storage vessel;
- a gas cylinder or cartridge designed to be disposed of when empty.

## Gas meter

Instrument that measures the volume of gas passing through it.

# 2. *Defining the end of the network and the installation*

## Network

The Network comprises interconnecting pipes which are downstream of a gasreception terminal, processing facility, storage facility or importing interconnector, and used for the conveyance of gas to consumers.

## End of the network

The end of the Network is the outlet of the emergency control valve (ECV) asdefined by relevant legislation and explained further in following section.

## Emergency control valve (ECV)

The ECV is a valve, not being an "additional emergency control valve" (AECV) (see Sub-Section) for shutting off the supply of gas in an emergency, intended for use by a consumer of gas and being installed at the end of aservice or distribution main. The outlet of the ECV terminates, and thus definesthe end of, the Network.

Note: The gas conveyor (which is, normally, a GT) has to agree the designation of the ECV which defines the end of the Network. For all "recommended gas supply arrangements", the ECV will be upstream of all components of the meter installation.

## Meter installation

(1) A meter installation includes a primary meter and any associated volume conversion sys-

tem, valve, filter, meter regulator or PRI, flexible connection, meter by-pass, interconnecting pipework, fitting and support.

(2) A meter installation commences at either:
- the outlet of the first common valve through which all the gas entering the meter installation will pass and which is upstream of the first meter regulator/PRI (including any filtration) upstream of the meter;
- in the case of a meter upstream of a regulator/PRI, or of an unregulated supply, the outlet-flange of the first common valve upstream of the primary filter(s) for the meter installation.

(3) A meter installation terminates at either:
- the outlet connection of the meter (if a meter outlet valve (MOV) is not fitted);
- the outlet of the meter outlet adapter if fitted;
- the outlet of the MOV (or outlet spool) if fitted;
- the outlet of the tee fitted downstream of the meter where a meter by-pass rejoins the pipework on the outlet of the meter;
- in the case of a meter upstream of a regulator/PRI, the outlet of the regulator/PRI outlet valve (PRIOV) or spool piece for a regulator by-pass;
- where a twin stream regulator/PRI is installed, the outlet of the tee where the two streams join;
- if provided, the outlet of the meter installation outlet valve (MIOV);
- in the case of a semi-concealed domestic meter with a flexible connection downstream of the meter, the outlet of the meter box outlet adapter, whichever is appropriate for the system.

## Additional emergency control valve (AECV)

An AECV is a valve, not being the ECV, for shutting off the supply of gas in an emergency, intended for use by a consumer of gas. An AECV may be located within either the meter installation or installation pipework and, as such, may not isolate all of the consumer's pipework or meter installation.

## Installation pipework

Installation pipework is any pipework or fitting from the outlet of the meter installation to points at which appliances/equipment are to be connected. It does not mean:
- a service (pipe) or distribution main or other pipeline;
- a pipe or fitting comprised in a gas appliance;
- a pipe or fitting within a meter installation;
- any valve attached to a storage container or cylinder.

# 3. *Natural gas*

## How natural gas is made

The natural gas that we use to heat our homes and our water comes from deep under the earth.

## Lesson 16 Introduction of Fuel Gas

The gas is found in layers of rock with tiny holes-the rock holds the gas like a sponge. To bring it to the surface, gas companies drill down hundreds of feet and pump into pipes.

## Components of natural gas

Natural gas is mostly methane. Methane ($CH_4$) is the simplest hydrocarbon molecule, with one atom of carbon and four of hydrogen.

## The efficient energy

We can know from the Fig. 16.1: With natural gas, 93 percent of the original energy that leaves the ground reaches your home. Only seven percent is lost in the drilling and distribution processes. Compare that to the electricity that is produced in this area by burning coal and natural gas. Only 28 percent of the original energy reaches your home as electricity. Seventy-two percent is lost in electricity's less efficient conversion and transmission methods.

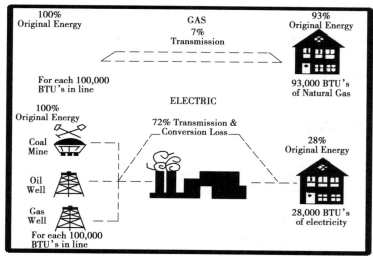

Fig. 16.1 **The efficient energy**

## Liquefied natural gas

**LNG**[①] stands for liquefied natural gas. It is natural gas cooled to roughly $-260$ °F at normal air pressure. It is odorless, non-toxic, non-corrosive and less dense than water. Essentially, it is the same natural gas people use to heat and cool their homes, only in a liquid state.

## Natural gas storage

Natural gas storage plays a vital role in maintaining the reliability of supply needed to meet the demands of consumers. Fig16.2 is the types of underground nature gas storage.

---

① 见 Part 3 Extension.

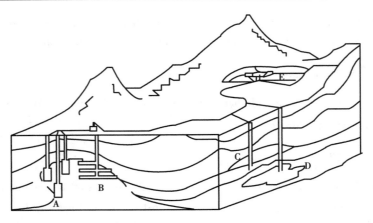

Fig. 16.2 **Underground storage**

A. Salt caverns; B. Mines; C. Aquifers; D. Depleted reservoirs; E. Hard-rock caverns

## Natural gas transport

Delivering natural gas is an enormous enterprise. The natural gas produced at remote sites must then be transported to consumers through very long transmission pipelines, vast storage reservoirs, and thousands of compressors that must all be efficient and reliable. Methods for establishing the nature gas transportation system are described in more detail in the next chapter.

## Natural gas application

(1) Natural gas blue flame heater

- Blue flame vent-free gas space heaters operate like your central heating system, gradually warming the air in the room.
- Low installation cost as this vent free unit requires no chimney or venting.
- No electricity required—ideal emergency heat.

(2) A natural gas fireplace

Natural gas fireplaces are structurally similar to wood-burning fireplaces. They are complete units that include a ceramic log set contained in a combustion chamber with a glass front. Their venting system eliminates the need for a traditional masonry chimney.

(3) Vehicles using natural gas

## Piping system design considerations

(1) Delivery conditions

Primary operating limitation of a pipe system is maximum allowable operating pressure (MAOP) of its weakest links. Weakest links is usually plastic pipe or components such as threads, flanges or valves. Regulatory-agency code provisions can also establish MAOP.

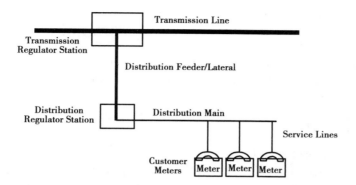

**Fig. 16.3 Simplified network schematic**

(2) Pipe size, **pressure**② & **material**③
- System design considerations:
    New business—new mains in new territory,
    Improvements—increase flow for growth,
    Replacements—corroded or damaged mains,
    Public improvements—highway widening etc. requiring facility relocation.
- Environmental considerations such as river crossings, refuges or sensitive areas,
- Pipe sizing is done with flow equations that calculate resistance to gas flow i. e. pressure drop, Longer pipe has more pressure drop at a given diameter. Tables exist based upon these flow, equations that given a desired flow rate and pressure drop give diameter & length.
- Pipe wall thickness determined by stresses:
    Primary—internal gas pressure,
    Secondary—thermal expansion/contraction etc.,
- Compression requirements:

Compression generally a transmission function or for distribution storage. Gas field compression is needed when field pressure is less than the pipeline. Industrial customers may require compression if they cannot take directly off the transmission line. Long transmission lines may require compressors spaced out to serve load.

- Pipeline routing:
    Route and location important cost drivers:
    Desire shortest possible route,
    Construction costs (pavement, traffic, etc.),
    Future customers,
    Future maintenance factors (corrosive soil, etc.),
    Government regulations,
    Construction barriers (rivers, rocky areas, major highways, etc.)—longer route may be cheaper.

---

②③ 见 Part 3 Extension.

**Keywords**: Natural gas, Fuel gas, Network, Metering

**Source from**:

[1] IGEM/G/4 Communication 1739. Definitions for the gas industry. Derbyshire: IGEM, 2009.
[2] IGEM/G/1. 3rd Communication 1733. Defining the end of the Network, a meter installation and installation pipework. Derbyshire: IGEM, 2008.

## Part 3　Extension

1. Liquefied natural gas(液化天然气)

液化天然气(简称LNG),主要成分是甲烷,被公认是地球上最干净的能源。无色、无味、无毒且无腐蚀性,其体积约为同量气态天然气体积的1/600,质量仅为同体积水的45%左右。其制造过程是先将气田生产的天然气净化处理,经一连串超低温液化后,利用液化天然气船运送。燃烧后对空气污染非常小,而且放出热量大。注意:液化天然气与液化气不同,液化气是指液化石油气。

2. Pipe pressure(管压)

城市管道是根据输气压力来分级的。这是由于燃气管道的气密性与其他管道相比,有特别严格的要求,漏气可能导致火灾、爆炸、中毒或其他事故。燃气管道中的压力越高,管道接头脱开或管道本身出现裂缝的可能性和危险性也越大。当管道内燃气的压力不同时,对管道材质、安装质量、检验标准和运行管理的要求也不同。

我国城市燃气根据输气压力一般分为:

低压燃气管道: $P \leqslant 0.01$ MPa

中压B燃气管道: $0.01$ MPa $< P \leqslant 0.20$ MPa

中压A燃气管道: $0.20$ MPa $< P \leqslant 0.40$ MPs

次高压B燃气管道: $0.4$ MPa $< P \leqslant 0.8$ MPa

次高压A燃气管道: $0.8$ MPa $< P \leqslant 1.6$ MPa

高压B燃气管道: $1.6$ MPa $< P \leqslant 2.5$ MPa

高压A燃气管道: $2.5$ MPa $< P \leqslant 4.0$ MPa

3. Pipe material(管材)

燃气工程管材的选择应根据输送介质的种类、设计压力、设计温度、工程所处的环境与焊接性能等因素确定。用于输送燃气的管材必须具有足够的机械轻度、优良的抗腐蚀性、抗震性、气密性、良好的韧性和焊接性能及易于连接等各项性能,同时各种管材的适应条件应与国家或行业现行标准和规程相适应。具体选择何种管材应经过技术经济比选。

## Part 4　Words and Expressions

| | |
|---|---|
| natural gas | 天然气 |
| liquefied petroleum gas(LPG) | 液化石油气 |
| combustible | *adj.* 易燃的,可燃的 |
| | *n.* 易燃物,可燃物 |
| component | *n.* 成分;零件;[数]要素;组分 |
| | *adj.* 组成的;合成的;构成的 |
| apparatus | *n.* 仪器,器械;机器;机构,机关 |

续表

| control valve (CV) | 控制阀 |
| filters | 过滤器 |
| slam-shut valve (SSVs) | 关断阀 |
| non-return valve (NRVs) | 止回阀 |
| gas meter | 煤气表 |
| conveyor | n. 运送者；传送者；传达者 |
| meter regulator | 计量器 |
| sponge | n. 海绵；海绵状物 |
|  | vt. (用海绵)擦拭 |
| methane | 甲烷 |
| salt cavern | 盐穴 |
| depleted reservoirs | 枯竭油藏 |
| hard-rock | 硬岩 |
| reservoir | n. 蓄水池；贮液器；储藏；蓄积 |
| ceramic | adj. 陶器的，与陶器有关的 |
|  | n. 陶瓷制品；陶瓷器；制陶艺术 |
| refuge | n. 避难；避难所；庇护者 |
|  | vt. 给予……庇护；接纳……避难 |

## Part 5　Reading

## Natural Gas Industry in China

### 1. *Value chain analysis of the natural gas industry*

The concept of value chain analysis as a generic business management tool was introduced by Porter (1985). Numerous studies on value chain analysis and value networks have been published since (e.g., Allee, 2003; and review in Weijermars, 2008). Industry stakeholders commonly benefit from a systemic value network analysis because it identifies key areas in the value network where constraints occur and opportunities for improvements arise. The global oil & gas industry is under considerable pressure to meet the world's demand for affordable and secure energy supply. Environmental concerns have intensified the interfuel competition and this battle can be prolonged in favour of optimum utility for the remaining global reserves of oil and gas.

Crucial in determining the cost of new transmission pipelines is the relative capital outlay on building, operating and maintenance cost of the pipe and of the compressor stations. It is necessary to account for the costs of construction of the line and the compressor stations, as well as the cost of running all the equipment. Models for efficient gas transmission focus on the two basic capital inputs forthe asset: pipes and compressors (e. g. Chenery, 1949). Compressor sare required to provide pressure for the gas transport, which decreases gradually due to frictional losses of energy when gas is moved along the pipe. The energy loss in the pipe due to friction in transmission is a decreasing function of pipe size. It follows that greater pipe diameters require less compressor capacity to pumpany given amount of gas over a specific distance. The conclusion is that cost optimization uses a substitution between pipes and compressors, based on calculations of energy loss and the effect of equipment's capacity size in reducing this loss (Robinson, 1972).

At present, Earth's proved ultimate recoverable reserves of conventional oil and gas amount to 2,000 Gbbls and 12,000 tcf, respectively (Laherrere and Wingert, 2008). Oil reserves from conventional sources can satisfy our oil consumption rate trend for the next 20 years and can be extended by unconventional oil reserves for at least an additional decade or more. Natural gas reserves last for another 80 to 100 years at the present consumptionrate trend. In 2008, the world consumed oil at a rate of 85 million bbls/d (30 Gbbls/y) and natural gas at a rate of 315 Bcf/d, corresponding to 115 tcf/y (DOE/EIA, 2009).

Ultimately, it is the demand of end-consumers who must pay for all engineering efforts by natural gas producers, shippers, transmission providers and retailers, which determines whether regional natural gas markets can develop successfully. Essentially, the physical value chain of natural gas that has been built around our planet is supported by a financial value chain, and vice versa.

## 2. *Origins and countermeasures for "gas famine" in china*

"Gas famine", occurred in early November of 2009 in some provinces of China, is due to three main reasons:

- the contradiction between supply and demand is the direct reason;
- the imperfect gas pricing system is the underlying reason;
- the incoherent development of natural gas industrial chain is the root cause.

Based on these points of view, a conclusion is drawn through an indepth analysis that the following counter measures should be taken to avoid the occurrence of "gas famine" and develop natural gas industry in a better way:

- the amount of gas supply should be increased by further exploration and exploitation of natural gas resources and importing more gas from other counties;
- gas transportation efficiency should be improved by well performed national gas networks resulting from gas pipeline projects;
- the capacity of gas supply in urgent cases should be enhanced by more matchedgas storage facilities;
- the reform of gas pricing system should be intensified.

## 3. "Gas famine" in china from the perspective of natural gas industrial chain

Three key factors help us well understand how "gas famine" broke out:

Particular stage of development accompanies the imbalance between supply and demand. Natural gas industrial chain has come to a stage of rapid development in the last decade, and is now even coming ahead of time to a stage of market-oriented resources division, which inevitably results in the imbalance between supply and demand.

Specific driving force triggers the imbalance between supply and demand. Competition to some degree exists in the up and middle stream of the industrial chain, pipeline construction projects are divided by state-owned oil companies and local governmental enterprises into trunk and branch lines, and low gas prices and pipeline gas connection charges stimulate end users especially urban gas dealers to try every means to attract users as many as possible.

Unique developing trajectory aggregates the imbalance between supply and demand. Although gas supply at the initial stage just met the need of agriculture and domestic use, now both economy and urbanization have stepped into a new level, resulting in a big demand for gas in the market, on the other hand, there is neither a complete plan of gas utilization nor good enough gas storage facilities and peak-shaving means.

Therefore, risk mitigation measures are suggested here:
- a full plan should be developed for the whole industrial chain;
- peak-valley adjustment mechanism should be gone through the whole chain;
- a gas pricing system should be reformed from the global perspective;
- a competitive system should be introduced gradually into the industrial chain by new reforms.

## 4. Safe supply of natural gas in china under global environment

To meet the rapidly increasing demand for natural gas indomestic energy market, the volume of pipeline gas and LNG will increase continually, so the degree of dependence on the overseas market is increasing gradually. Supply disruption brings risk in gassupply, so a big challenge is facing up with safe and stable supply of natural gas. Based on the current status and developing trend of gas supply in the Asia-Pacific regions, we should focus on the situation and challenges of gas supply in China, and the framework and some proposals for safety and risk management in gas supply are presented here as follows:

Government should concentrate on energy political and foreign affairs, complete the related legislation and regulations of gas supply, and work out natural gas pricing mechanism;

Gas supply companies should make more efforts in exploring and recovering domestic gas resources, introducing steadily overseas gases as supplementary energy sources, and speed up the construction of infrastructure facilities like the pipelines, gas storages, etc.;

Industrial and domestic gas consumers should contribute to peak-shaving especially in urban gas supply by building some facilities, with which gas supply will be guaranteed especially in cold seasons and some other urgent cases.

**Source from:**

[1] Ruud Weijermars. Journal of Natural Gas Science and Engineering. Elsevier, 2010, 2:86-104.

[2] Dong Xiucheng. Origins and countermeasures for "gas famine" in China. Natural Gas Industry, 2010, 30:119 – 122.

## Part 6　Exercises and Practices

### 1. *Discussion*

What kind of material is often used to produce gas pipes?

### 2. *Translation*

(1) System comprising a distribution main/service (pipe) emergency control valve, meter installation and installation pipework and any additional emergency control valve (AECV) to supply a consumer's appliance.

(2) The ECV is a value, not being an "additional emergency control valve" (AECV) (see Sub-Section) for shutting off the supply of gas in an emergency, intended for use by a consumer of gas and being installed at the end of a service of distribution main.

(3) A meter installation includes a primary meter and any associated volume conversion system, valve, filter, meter regulator or PRI, flexible connection, meter by-pass, interconnecting pipework, fitting and support.

(4) The natural gas produced at remote sites must then be transported to consumers through very long transmission pipelines, vast storage reservoirs, and thousands of compressors that must all be efficient and reliable.

(5) Primary operating limitation of a pipe system is maximum allowable operating pressure (MAOP) of its weakest links.

### 3. *Writing*

Please explain what should be taken into consideration when we choose the size of pipe?

# Lesson 17

# Gas Transportation Process

## Part 1  Preliminary

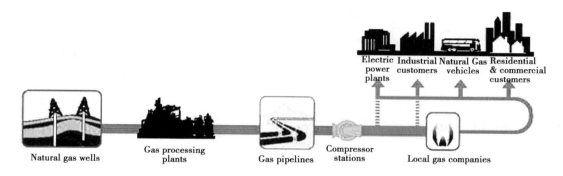

Fig. 17.1  **Natural gas distribution flow chart**

The pure natural gas extracted from gas fields is usually transported through gas pipelines to towns and industrial areas far from gas fields. Oilfield gas and artificial gas with large output can also be delivered to distant gas areas in the same way. Figure 17.1 simply shows the flow of gas transmission system. Natural gas transportation systems generally consist of mine gathering and transportation systems, gas processing plants, gas pipeline starting stations, gas pipelines, intermediate compressor stations, management and maintenance stations, communication and remote control facilities, cathodic protection stations, and gas distribution stations and so on. Due to differences in air source pressure, type, temperament and transport distance, the station settings of gas transportation systems are also different.

## Part 2  Text

燃气长距离输送系统的作用是将产气地的气源(天然气、油田气及人工煤气)输送到远离产地的使用地(城镇和工业园)。矿场集输系统的作用是对天然气进行节流减压、初次分离和计量。输气干线的起点站一般是调压计量站,主要任务是保持输气压力平稳,对燃气压力进行自动调节、计量以及除去燃气中的液滴和机械杂质。输气干线或支线的终点是燃气分配站,作为城镇或工业区分配管网的气源站,经过除尘、调压、计量和加臭后进入城镇或工业区。

After the gas field exploration, development and mining, the natural gas gathering and transportation system becomes a very important production process. Firstly, this part introduces the natural gas transportation system. Then it provides advice on the potential reasons for cessation of supplies to commercial and, in particular, industrial gas users and the actions necessary to be taken in a Natural Gas supply emergency.

## 1. The Natural gas transportation system

Broadly speaking, the transportation starts from the wellhead, including collecting the natural gas through pipelines and treating the gas through processing plants. After **purification**[①], the gas will become the qualified commodities, which can be send to the user and participate the whole production process.

### The gas well

Generally, gas wells will be equipped with the drill device locally. Simple gas wells only have the "Christmas tree" device. Facing the need to deal with the single well, it is necessary to set up a set of pressure regulation, separation and measuring equipment. After the **throttling step-down**[②], the free water, condensate oil and mechanical impurities will be removed from the gas in the segregator. Then, the gas goes into the transmission lines after metering.

### Gas-gathering stations

Generally, wells are connected to gas gathering stations by the pipeline. Natural gas is throttled, separated (oil, water, mechanical impurities) and metered in the gathering station, and then enter the gas gathering pipelines. The Gas gathering station is divided into two forms: the normal temperature gas station and the low-temperature gas gathering station and the latter is more complex than the former.

### Gas-processing plants

With pooled pretreatment of natural gas containing $H_2S$ and other impurities in the gas gathering station, the gas still need to enter the gas processing plant. In the processing plant, the gas will be taken off hydrogen sulfide, carbon dioxide, condensate and water, then gas quality reaches the tube-transport and commercial natural gas quality standards.

### Compressor stations

Compressor stations are divided into mine gas compression stations and trunk line gas compression stations of these two types. In the gas field development period (or the low pressure gas field), when the well head pressure cannot meet the production and distribution required pressure, you have to set the compressor station. In addition, the flow of natural gas in the pipeline resulting in the pressure decreased. To ensure the pipe-conveying ability does not decline, the compressor station must be set in a certain position on the pipeline.

---

①② 见 Part 3 Extension.

## Regulating-metering stations (Gas Distribution Stations)

Generally the regulating-metering station set on the starting point of the gas branch line or the end of the trunk line, which (maybe more than one) can also set in a location on the trunk line. Its task is to allocate the gas to the user.

## Gathering lines & transmission lines

Gathering lines include the connection pipe between gas wells and gas gathering stations, and the pipe between gas gathering stations and treatment plants.

From the gas purification plant to the distant gas users, the pipeline called transmission lines. When transmission lines cross the railways, highways, rivers and valleys, crossing projects are necessary.

## Go-devil stations[3]

For the removal of the fluid and dirt within the pipe to improve the transmission capacity of the pipeline, the pipeline pigging station often set on the trunk line and the gas gathering trunk.

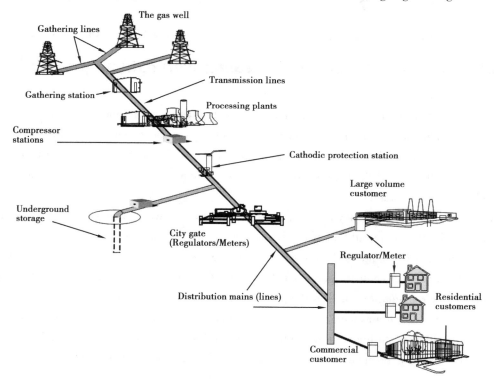

Fig. 17.2　The natural gas transportation system

---

③　见 Part 3 Extension.

## Cathodic protection stations

In order to prevent and delay the trunk line buried in the soil within the electrochemical corrosion, every certain distance set a cathodic protection station on the trunk line.

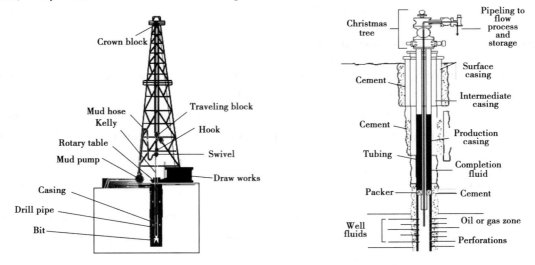

Fig. 17.3  Drilling rig          Fig. 17.4  A completed well

## 2. *Advices for big consumers in dealing with natural gas supply emergencies*

This part provides advice on the potential reasons for cessation of supplies to commercial and, in particular, industrial gas users and the actions necessary to be taken in a Natural Gas supply emergency. At the same time, this part aims to give simple advice to Natural Gas consumers on the consequences of the loss of their gas supply and technical requirements to ensure continued safe operation of consumers' plant.

With appropriate contingency planning, gas consumers may avoid being exposed to greater costs and commercial risks than would otherwise be expected. Minimizing the impacts of potential Natural Gas supply emergencies requires the co-operation of everyone in the industry. Only by working together can the quality of contact information be improved. This is a vital part of the industry emergency process for reducing demand on the gas network quickly and safely.

In the event of such an emergency, it is vital that Large End Users stop using gas as soon as can be safely achieved, to limit the disruption to smaller consumers and to safeguard the Network. However, the need for competent persons, planning and safe procedures is of paramount importance.

### Planning for cessation of gas supply

(1) Causes of cessation of gas supply

The cessation of supplies to users of Natural Gas may be caused by:

- damage to, or failure of, the local gas supply network;
- failure of part of the national transmission system;
- insufficient supplies to meet gas demand.

In such cases, large end users may be required to implement immediate or phased cessation of their gas supply.

*Note: Large users will be directed to cease using gas ahead of priority users and domestic users. The proximity of a premises' supply to a network terminal does not imply that gas supply will be secure under emergency conditions.*

Under any of the above circumstances, the user may be contacted by the gas transporter, or shipper or supplier acting on the transporter's behalf, and directed to cease using gas. If a user fails to comply with the direction to cease consumption of gas and the gas transporter attends to isolate the user's supply, this will normally be done at the Emergency Control Valve (at the primary meter). The user's co-operation with the gas transporter in an emergency is vital if the safety of the gas supply network is to be assured.

(2) Planning for cessation of gas supply

There are a number of important steps which the user should take in planning for a cessation of supply:

- assess the risks to site and business operations due to the cessation of gas supplies.

*Note: This may include business continuity issues.*

- establish who is the gas transporter, gas shipper and gas supplier and adopt a rigorous procedure to ensure their details are kept up-to-date.

*Note: Details of the gas transporter and supplier are shown on the gas supply invoice. The gas supplier will have details of the relevant gas shipper.*

- prepare an operating plan for when no gas is available, including the availability/operability of existing alternative site energy sources, and potential damage to plant or processes e.g. caused by freezing of water systems.
- identify the key contact points within the user's company to respond to the direction to curtail supplies. It is vital that up-to-date details of these contacts are supplied to the gas supplier, including "out of hours" details [24 hours/365 days].
- ensure that competent resources are available to cease the use of gas as directed.
- ensure competent resources are available for the restoration of gas systems and the start-up of associated plant and equipment when the gas supply emergency has finished.

*Note: Other Large End Users may be similarly affected and, therefore, if the required competency to restore gas supplies is not available "in-house", it is worth bearing in mind that the services of specialized contractors may be stretched at this time.*

- on a regular basis, establish and test the procedures for the controlled shutdown and start-up (re-commissioning) of plant and equipment.
- if the risk assessment indicates that significant health and safety implications may exist due to cessation of supply, the user may contact the gas supplier.
- ensure it is known who will provide the information that the gas supply emergency has fin-

ished i. e. will it be the transporter, the shipper, the supplier or another party?

## Shutting down plant and equipment

(1) When a direction to cease consumption is received then the user should try to do this by shutting down gas appliances rather than turning off the gas at the Emergency Control Valve (at the primary meter) or at other section isolation valves. The primary consideration is to ensure continuing safety while, if reasonably practicable, maintaining a positive gas pressure i. e. greater than atmospheric pressure, within the pipework system. Turning off the supply at the Emergency Control Valve or at any other section isolation valve may lead to the complete loss of pressure within the installation pipework downstream of that valve. This may also lead to the inability to ignite alternative fuels and require purging, testing and commissioning upon resumption of supply. These activities could involve considerable expense and delay. Furthermore, in some cases, the steps to re-commission will require a written methodology and the use of appropriate purge gas. There will also be a need for staff with the appropriate level of competence to test and purge the system.

Note: *There may occasions where the user suspects a failure of the gas supply network before being contacted by the gas transporter, or shipper or supplier. In such cases, each appliance should immediately be turned off.*

(2) If the user has the ability to use alternative fuels, it is important to ensure that all associated equipment can be readily switched to the alternative fuel at short notice.

Note 1: *It may be necessary to have either a standby LPG or bottled Natural Gas supply for the ignition system.*

Note 2: *In a gas supply emergency, little or no advance notice may be given of the cessation of gas supply, so it is important to remember that adequate stand by fuel storage or heated oil systems may be needed to permit rapid change over to oil firing at any time and not just under peak winter conditions.*

## Re-starting equipment upon restoration of gas supply

When the gas supply emergency is finished, you will be notified by the gas transporter, shipper or supplier.

(1) Gas supply locally isolated at appliances

If appliances are isolated at their local isolation valve, it should be possible to light the appliances one at a time. Check in each case that they appear to be operating correctly. Site staff should understand and be sufficiently competent to attempt to relight appliances. They should also have the operating and fault finding instructions to hand. If there is any doubt about the staff competency or if appliances do not appear to be operating correctly, assistance should be seek from a competent person such as the manufacturer/servicing agent.

(2) Gas supply isolated at the emergency control valve or section isolation valves

In this instance, it is essential to first apply a test for gas tightness of the existing pipework system before purging takes place. This must be performed by persons holding the appropriate competency.

**Keywords**: Transportation, Supply emergency, Gas supply, Gas company

**Source from**:
IGEM, Gas Legislation Guidance IGE/GL/9 Communication 1724. Guidance for large gas consumers in dealing with Natural Gas supply emergencies. Derbyshire: IGEM, 2006.

## Part 3　Extension

1. Purification(净化)

从气井中采集到的天然气中含有水、砂、酸气等许多固液杂质。这些杂质的存在将消弱管线的输送能力(固性杂质容易造成设备仪表损坏;液态、固性水化物容易阻塞管道;酸性气体溶于水腐蚀燃气管道等),同时也降低了燃气本身的热值。因此,在进入长输管线之前,必须使用物理、化学及生化等手段对燃气进行净化。

2. Throttling step-down(节流降压)

流体在管道内流动,遇到突然变窄的断面,由于存在阻力使流体的压力降低的现象称为节流。"节流降压"实质是采用节流的方式,降低管道中流体的压力、温度。在天然气输送工艺中使用节流效应主要是为了降低天然气的温度,从而使天然气中的一部分水分实现冷凝分离,是天然气净化工艺中的一种物理净化技术。

3. Go-devil stations(清管器站)

清管器是由气体、液体或管道输送介质推动,用以清理管道的专用工具。对于运营中的天然气管线、清管器的作用:清除管线内部积水、轻质油、甲烷水合物、氧化铁、碳化物粉尘、二硫化碳、氢硫酸等腐蚀性物质;降低腐蚀性物质对管道内壁的腐蚀损伤;重新明确管线走向;检测管线变形;检查沿线阀门完好率;减小工作回压。

## Part 4　Words and Expressions

| | |
|---|---|
| cessation | *n.* (暂时)停止,休止,中断 |
| wellhead | *n.* 水源,泉源 |
| purification | 净化 |
| commodity | *n.* 商品;日用品;有价值的物品;有利 |
| the gas well | 气井 |
| christmas tree | 采油树 |
| throttling step-down | 节流降压 |
| mechanical impurity | 机械杂质 |
| gas-gathering station | 集气站 |
| pooled | *adj.* 合并的 |
| | *v.* 集中……共同使用,共用 |
| compressor station | 加压站 |
| trunk line | 主干线 |
| regulating-metering stations | 调压计量站 |
| go-devil station | 清管器站 |

续表

| pigging | n. 生铁；通井 |
| cathodic protection station | 阴极保护站 |
| contingency | n. 意外事故，偶发事件；可能性 |

## Part 5  Reading

## Some Issues of Natural Gas Gathing & Transportation

### 1. Long-distance gas pipeline design in new situation

Design project management of long-distance oil and gas pipelines is a complex system management, pipeline construction in recent years presents such new situation as taking the design as the principal thing, routing coordination throughout construction period, complex external relations and big differences between local requirements, which makes the requirements for pipeline construction more strict and its execution more difficult than before.

Current gas pipeline construction in China has such characteristics as numerous projects under construction, wide regional distribution and short time limit for a project and its design schedule and quality and application of standardization achievements are critical to successful completion of a project.

Taking feasibility study, preliminary design, engineering, procurement and construction and project management contracting into consideration, some discussions should comparatively analyze the differences between traditional designs and new designs under new situation and discuss some specific practices for completing design with high quality and efficiency.

### 2. Natural gas purification process design and their optimization

Up to now, amine is still widely used in sour gas desulfurization and decarbonization at home and abroad and has been used for more than 50 years in China. Blocking solution may occur in desulfurization and decarbonization unit during production and result in serious capacity decline of desulfurization and decarbonization and not qualified gas purification degree. Solution foaming may result in entrainment and a large number of amine solution may flow away with air stream, which will lead to sharp increase in solution loss and serious economic losses.

We should summarize the successful experience in natural gas purification process design and production, to propose such key points as process design of desulfurization and decarbonization unit and key equipment selection. And we should describe some considerations during design of such systems in desulfurization and decarbonization unit as feed gas separation system, amine regeneration system, amine filtration and inert gas protection system, amine and acid gas cooling system, lean amine turbocharger and circulation pump, and analyze design of such key equipment in purification process as columns, vessels, cold and heat exchanger and pumps.

## 3. Underground natural gas storage with salt caverns

During operation of underground natural gas storage with salt caverns, thermal exchange between gas in salt cavern and surrounding salt bed, continuous cycle of injection and recovery and Joule-Thomson effect will affect stability and safety of salt cavern, make gas form hydrate, which will obstruct gas transportation. So it is necessary to research operation characteristic of salt cavern gas storage and to select the best operation scheme.

We should discuss, based on correlative technical literatures and research fruits at home and abroad, the operation characteristic of salt cavern natural gas storage and critical technology in its operation and some suggestions should be put forward for future domestic studies on salt cavern natural gas storage.

## 4. Protection of anticorrosive coating in gas pipe line

In recent years, directional drilling crossing technique has been widely used in China. In order to avoid external anticorrosive coating in gas pipe line from being damaged during directional drilling crossing, the feasible protection scheme should be put forward after analysis on present drilling crossing construction techniques, corrosion control performance of external anticorrosive coating and actual conditions in construction site, which could be efficient in actual operation.

## 5. The safety assessment on natural gas gathering & transportation station—application of entropy technology

Because natural gas gathering and transportation system has such features as multiple stations, long pipeline and wide range and many risky factors exist in production, it is necessary to reflect overall safety conditions of the system accurately.

With the application of entropy technology, we should establish an index system for multistage fuzzy comprehensive assessment, apply fuzzy mathematics theory, adopt the method for objective weight calculation in information theory and propose the two-stage fuzzy comprehensive assessment method. This method is used for comprehensive assessment on safety conditions in some gas gathering and transportation station and the conclusion of "Better" is obtained. This method can not only find out deficiencies existing in gathering and transportation station, but also provide guidance for establishing safety countermeasures and safeguards.

**Source from:**

[1] Yang Fan. Discussion on Long-distance Gas Pipeline Design in New Situation. *Natural Gas and Oil*, 2012, 30: 14-16.

[2] Liang Junyi. Key Points of Natural Gas Purification Process Design and Their Optimization. *Natural Gas and Oi*, 2012, 30:32-35.

[3] Wu Xuehong. Operation Management of Underground Natura Gas Storage with Salt Caverns. *Natural Gas and*

Oil, 2008, 26:1-4.
[4] Wang Tao. Application of Entropy Technology in Safety Assessment on Natural Gas Gathering and Transportation Station. *Natural Gas and Oil*, 2012, 30:10-13.
[5] CIBSE Guide B. Heating, ventilation, air conditioning and refrigeration: Section 1 Heating. London: CIBSE Publications, 2005.
[6] IGEM/G/4 Communication 1739. Definitions for the gas industry. Derbyshire: IGEM, 2009.
[7] IGEM/G/1. 3rd Communication 1733. Defining the end of the Network, a meter installation and installation pipework. Derbyshire: IGEM, 2008.
[8] Dong Xiucheng. Origins and countermeasures for "gas famine" in China. Natural Gas Industry, 2010, 30: 119-122.
[9] IGEM, Gas Legislation Guidance IGE/GL/9 Communication 1724. Guidance for large gas consumers in dealing with Natural Gas supply emergencies. Derbyshire: IGEM, 2006.

## Part 6    Exercises and Practices

### 1. Discussion

(1) During directional drilling crossing, what should we pay attention to?

(2) What measures should be taken, by the large end user, when natural gas supply emergencies happen?

### 2. Translation

(1) After purification, the gas will become the qualified commodities, which can be send to the user and participate the whole production process.

(2) In the processing plant, the gas will be taken off hydrogen sulfide, carbon dioxide, condensate and water, then gas quality reaches the tube-transport and commercial natural gas standards.

(3) In the gas filed development period (or the low pressure gas filed), when the well head pressure cannot meet the production and distribution required pressure, you have to set the compressor station.

(4) In the event of such an emergency, it is vital that Large End Users stop using gas as soon as can be safely achieved, to limit the disruption to smaller consumers and to safeguard the Network.

(5) Current gas pipeline construction in China has such characteristics as numerous project under construction, wide regional distribution and short limit for a project, its design schedule and quality and application of standardization achievements are critical to successful completion of a project.

### 3. Writing

Please explain the role of cathodic protection station.

# Lesson 18
# Building Energy Efficiency

## Part 1　Preliminary

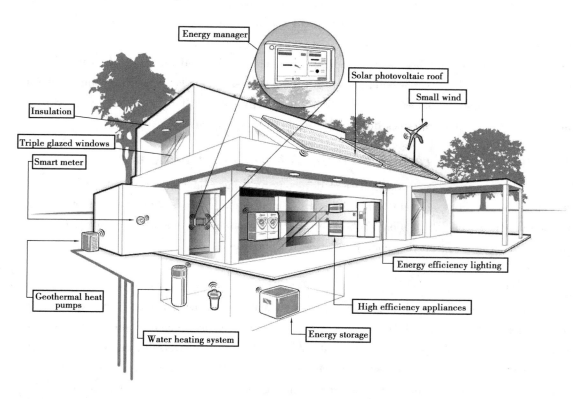

Fig. 18.1　**Low energy house**

　　A low-energy house is a building that has a very high energy performance and nearly zero or a very low amount of energy required. The low energy design of buildings has been considered an important goal both to encourage resource efficiency and to reduce the potential for global climate change associated with the consumption of fossil fuels. Broadly speaking, the design involves two strategies-minimizing the need for energy use in buildings (especially for heating and cooling) through EEMs (energy-efficient measures) and adopting RETs (renewable energy and other technologies) to meet the remaining energy needs. EEMs include building envelopes, internal conditions, and building services systems; RETs cover photovoltaic or building-integrated photovoltaic, wind turbines, solar thermal (solar water heaters), heat pumps, and district heating and cooling. These include life-cycle cost and environmental impacts, climate change and social policy issues.*

## Part 2　Text

　　建筑节能的直接目的是提高建筑使用过程中的能源利用效率。我国建筑节能承担着双重任务:一是改善建筑环境,提高居住水平;二是提高建筑的能源利用效率,节能减排。建筑领域的节能主要包括建筑围护结构的节能、建筑设备系统的性能优化、可再生能源在建筑中的使用、建筑系统的优化控制和运行管理。

　　Most energy used in buildings is from nonrenewable resources, the cost of which historically has not considered replenishment or environmental impact. Thus, consideration of energy use in design has been based primarily on economic advantages, which are weighted to encourage more rather than less use.

　　As resources become less readily available and more exotic, and replenishable sources are investigated, the need to operate buildings effectively using less energy becomes paramount. Extensive study since the mid-1970s has shown that **building energy use**①can be significantly reduced by applying the fundamental principles discussed in the following sections.

### 1. *Energy ethic: resource conservation design principles*

　　The basic approach to energy-efficient design is reducing loads (power), improving transport systems, and providing efficient components and "intelligent" controls. Important design concepts include understanding the relationship between energy and power, maintaining simplicity, using self-imposed budgets, and applying energy-smart design practices.

### 2. *Energy and power*

　　From an economic standpoint, more energy-efficient systems need not be more expensive than less efficient systems. Quite the opposite is true because of the simple relationship between energy and power, in which power is simply the time rate of energy use (or, conversely, energy is power times time). Power terms such as kilowatts are used in expressing the size of a motor, chiller, boiler, or transformer. Generally, the smaller the equipment, the less it costs. Other things being equal, as smaller equipment operates over time, it consumes less energy. Thus, in designing for energy efficiency, the first objective is always to reduce the power required to the bare minimum necessary to provide the desired performance, starting with the building's heating and cooling loads a power term, in kilowatts and continuing with the various systems and subsystems.

### 3. *Simplicity*

　　Complex designs to save energy seldom function in the manner intended unless the systems are continually managed and operated by technically skilled individuals. Experience has shown that

---

\* 注:来源于维基百科
① 见 Part 3 Extension.

long-term, energy-efficient performance with a complex system is seldom achievable. Further, when complex systems are operated by minimally skilled individuals, both energy efficiency and performance suffer. Most techniques discussed in this chapter can be implemented with great simplicity.

## 4. *Self-Imposed budgets*

Just as an engineer must work to a cost budget with most designs, self-imposed power budgets can be similarly helpful in achieving energy-efficient design. The series of Advanced Energy Design Guides from ASHRAE are a source for guidance on achievable design budgets. For example, the following are possible categories of power budgets for a mid-rise office building:

- Installed lighting (overall) $W/m^2$;
- Space sensible cooling $W/m^2$;
- Space heating load $W/m^2$;
- Electric power (overall) $W/m^2$;
- Thermal power (overall) $W/m^2$;
- Hydronic system head kPa;
- Water chiller (water-cooled) kW/kW cooling;
- Chilled-water system auxiliaries kW/kW cooling;
- Unitary air-conditioning systems kW/kW cooling;
- Annual electric energy $MJ/(m^2 \cdot yr)$;
- Annual thermal energy $kJ/(m^2 \cdot yr \cdot K \cdot day)$.

As the building and systems are designed, all decisions become interactive as each subsystem's power or energy performance is continually compared to the budget.

## 5. *Design process for energy-efficient projects*

Consider energy efficiency at the beginning of the building design process, because energy-efficient features are most easily and effectively incorporated at that time. Seek the active participation of all members of the design team, including the owner, architect, engineer, and often the contractor, early in the design process. Consider building attributes such as building function, form, orientation, window/wall ratio, and HVAC system types early in the process, because each has major energy implications. Identify meaningful energy performance benchmarks suited to the project, and set project-specific goals. Energy benchmarks for a sample project are shown in Table 18.1. Consider energy resources, on-site energy sources, and use of renewable energy, credits, or carbon offsets to mitigate environmental impacts of energy use.

Tab 18.1 Example benchmark and energy targets for university research laboratory

| Bullding area, m² | Gross 15 793 | LIt/ Conditioned 10 266 | | | | | | |
|---|---|---|---|---|---|---|---|---|
| Electric | Electricity for Lighting | Electricity for Ventilation (Fans) | Electricity for In-Building Pumps | Electricity for Plug Loads | Electricity for Unidentified Loads | Total Electricity | Cogenerated Electricity | NGrid Electrictiy |
| Design load, W/m² Peak demand, W/m² Peak demand, kW (Projected submetered peak) | 5.60 4.52 71 | 5.38 5.38 85 | 6.46 4.52 72 | 10.4 7.86 124 | — 0.001 7 20 | 28.0 22.3 372 | — — — | |
| Annual consumption, kWh/yr (Projected submetered reading) | 218 154 | 346 598 | 191 245 | 891 503 | 175 200 | 1 823 000 | 966 000 | 857 000 |
| Annual use index goal, kWh/yr | 1.28 | 2.04 | 1.12 | 5.24 | 1.03 | 10.72 | | |
| Annual use index goal, site MJ/m² gross · yr | 4 378 | 6 956 | 3 838 | 17 893 | 3 516 | 36 583 | | |
| Annual use index, kWh/m² gross · yr * | 27.0 to 35.7 | 48.2 to 74.0 | included elsewhere | 47.3 to 61.0 | NA | 158.7 to 192.8 | | |
| Annual use index, site MJ/m² gross · yr * | 97.3 | 173.6 | — | 170.1 | — | 571.1 to 694.0 | | |

* From Labs 21 program of U.S Environmental Protection Agency (EPA) and U.S. Department of Energy (DOE). See http://www.epa.gow/lab21gov/index.htm.

Address a building's energy requirements in the following sequence:

• Minimize the impact of the building's functional requirements by analyzing how the building relates to its external environment. Advocate changes in building form, aspect ratio, and other attributes that reduce, redistribute, or delay (shift) loads. The load calculation should be interactive so that the effect of those factors can be seen immediately.

• Minimize loads by analyzing external and internal loads imposed on the building energy-using subsystems, both for peak-and part-load conditions. Design for efficient and effective operation off-peak, where the majority of operating hours and energy use typically occurs.

• Maximize subsystem efficiency by analyzing the diversified energy and power requirements of each energy-using subsystem serving the building's functional requirements. Consider static and dynamic efficiencies of energy conversion and energy transport subsystems, and consider opportunities to reclaim, redistribute, and store energy for later use.

• Study alternative ways to integrate subsystems into the building by considering both power and time components of energy use. Identify, evaluate, and design each of these components to control overall design energy consumption.

## 6. *Building energy use elements*

### Envelope

Control thermal conductivity by using insulation (including movable insulation), thermal mass, and/or phase-change thermal storage at levels that minimize net heating and cooling loads on a time-integrated (annual) basis.

• Minimize unintentional or uncontrolled thermal bridges, and include them in energy-related calculations because they can radically alter building envelope conductivity. Examples include wall studs, balconies, ledges, and extensions of building slabs.

• Minimize infiltration so that it approaches zero. (An exception is when infiltration provides the sole means of ventilation, such as in small residential units.) This minimizes fan energy consumption in pressurized buildings during occupied periods and minimizes heat loss (or unwanted heat gain, in warm climates) during unoccupied periods. In warm, humid climates, a tight envelope also improves indoor air quality. Reduce infiltration through design details that enhance the fit and integrity of building envelope joints in ways that may be readily achieved during construction (e.g., caulking, weatherstripping, vestibule doors, and/or revolving doors), with construction meeting accepted specifications.

• Consider operable windows to allow occupant-controlled ventilation. This requires careful design of the building's mechanical system to minimize unnecessary HVAC energy consumption, and building operators and occupants should be cautioned about improper use of operable windows. CIBSE (2005) provides comprehensive design considerations for natural ventilation.

• Strive to maintain occupant radiant comfort regardless of whether the building envelope is designed to be a static or dynamic membrane. Design opaque surfaces so that average inside surface temperatures remain within 3 K of room temperature in the coldest anticipated weather (i.e., winter design conditions) and so that the coldest inside surface remains within 14 K of room temperature (but always above the indoor dew point). In a building with time-varying internal heat generation, consider thermal mass for controlling radiant comfort. In the perimeter zone, thermal mass is more effective when it is positioned inside the envelope's insulation.

• Effective control of solar radiation is critical to energy-efficient design because of the high level of internal heat production already present in most commercial buildings. In some climates, lighting energy consumption savings from daylighting techniques can be greater than the heating and cooling energy penalties that result from additional glazed surface area required, if the building envelope is properly designed for daylighting and lighting controls are installed and used. (In other climates, there may not be net savings.) Daylighting designs are most effective if direct solar beam radiation is not allowed to cause glare in building spaces.

• Design transparent parts of the building envelope to prevent solar radiant gain above that necessary for effective daylighting and solar heating. On south-facing facades (in the northern hemisphere), using low shading coefficients is generally not as effective as external physical shading devices in achieving this balance. Consider low-emissivity, high-visible-transmittance glazings for effective control of radiant heat gains and losses. For shading control, judicious use of vegetation may block excess gain year-round or seasonally, depending on the plant species chosen.

## Lighting

Lighting is both a major energy end use in commercial buildings (especially office buildings) and a major contributor to internal loads by increasing cooling loads and decreasing heating loads. Design should meet both the lighting functional criteria of the space and minimize energy use. IESNA (2000) recommends illuminance levels for visual tasks and surrounding lighted areas.

## Other loads

• Minimize thermal impact of equipment and appliances on HVAC systems by using hoods, radiation shields, or other confining techniques, and by using controls to turn off equipment when not needed. Where practical, locate major heat-generating equipment where it can balance other heat losses. Computer centers or kitchen areas usually have separate, dedicated HVAC equipment. In addition, consider heat recovery for this equipment.

• Use storage techniques to level or distribute loads that vary on a time or spatial basis to allow operation of a device at maximum (often full-load) efficiency.

## HVAC system design[②]

• Consider separate HVAC systems to serve areas expected to operate on widely differing operating schedules or design conditions. For instance, systems serving office areas should generally be separate from those serving retail areas.

• Arrange systems so that spaces with relatively constant, weather-independent loads are served by systems separate from those serving perimeter spaces. Areas with special temperature or humidity requirements (e.g., computer rooms) should be served by systems separate from those serving areas that require comfort heating and cooling only. Alternatively, provide these areas with supplementary or auxiliary systems.

• Sequence the supply of zone cooling and heating to prevent simultaneous operation of heating and cooling systems for the same space, to the extent possible. Where this is not possible because of ventilation, humidity control, or air circulation requirements, reduce air quantities as much as possible before incorporating reheating, recooling, or mixing hot and cold airstreams. For example,

---

② 见 Part 3 Extension.

if reheat is needed to dehumidify and prevent overcooling, only ventilation air needs to be treated, not the entire recirculated air quantity. Finally, reset supply air temperature up to the extent possible to reduce reheating, recooling, or mixing losses.

• Provide controls to allow operation in occupied and unoccupied modes. In the occupied mode, controls may provide for a gradually changing control point as system demands change from cooling to heating. In the unoccupied mode, ventilation and exhaust systems should be shut off if possible, and comfort heating and cooling systems should be shut off except to maintain space conditions ready for the next occupancy cycle.

• In geographical areas where diurnal temperature swings and humidity levels permit, consider judicious coupling of air distribution and building structural mass to allow nighttime cooling to reduce the requirement for daytime mechanical cooling.

• High ventilation rates, where required for special applications, can impose enormous heating and cooling loads on HVAC equipment. In these cases, consider recirculating filtered and cleaned air to the extent possible, rather than 100% outside air. Also, consider preheating outside air with reclaimed heat from other sources.

## HVAC equipment selection

• To allow HVAC equipment operation at the highest efficiencies, match conversion devices to load increments, and sequence the operation of modules. Oversized or large-scale systems should never serve small seasonal loads (e. g. , a large heating boiler serving a summer-service water-heated load). Include specific low-load units and auxiliaries where prolonged use at minimal capacities is expected.

• Select the most efficient (or highest-COP) equipment practical at both design and reduced capacity (part-load) operating conditions.

• When selecting large-power devices such as chillers (including their auxiliary energy burdens), economic analysis of the complete life-cycle costs should be used.

• Keep fluid temperatures for heating equipment devices as low as practical and for cooling equipment as high as practical, while still meeting loads and minimizing flow quantities.

## Energy transport systems

Energy should be transported as efficiently as possible. The following options are listed in order of efficiency, from the lowest energy transport burden (most efficient) to the highest (least efficient):

(1) Electric wire or fuel pipe;
(2) Two-phase fluid pipe (steam or refrigerant);
(3) Single-phase liquid/fluid pipe (water, glycol, etc.);
(4) Air duct.

Select a distribution system that complements other parameters such as control strategies, storage capabilities, conversion efficiency, and utilization efficiency.

The following specific design techniques may be applied to thermal energy transport systems:

(1) Steam Systems

- Include provisions for seasonal or non-use shutdown.
- Minimize venting of steam and ingestion of air, with design directed toward full-vapor performance.
- Avoid subcooling, if practical.
- Return condensate to boilers or source devices at the highest possible temperature.

(2) Hydronic Systems

- Minimize flow quantity by designing for the maximum practical temperature range.
- Vary flow quantity with load where possible.
- Design for the lowest practical pressure rise (or drop).
- Provide operating and idle control modes.
- When locating equipment, identify the critical pressure path and size runs for the minimum reasonable pressure drop.

(3) Air Systems

- Minimize airflow by careful load analysis and an effective distribution system. If the application allows, supply air quantity should vary with sensible load (i.e., VAV systems). Hold the fan pressure requirement to the lowest practical value and avoid using fan pressure as a source for control power.
- Provide normal and idle control modes for fan and psychrometric systems.
- Keep duct runs as short as possible, and keep runs on the critical pressure path sized for minimum practical pressure drop.

## Power distribution

- Size transformers and generating units as closely as possible to the actual anticipated load (i.e., avoid oversizing to minimize fixed thermal losses).
- Consider distribution of electric power at the highest practical voltage and load selection at the maximum power factor consistent with safety.
- Consider tenant submetering in commercial and multifamily buildings as a cost-effective energy conservation measure. (A large portion of energy use in tenant facilities occurs simply because there is no economic incentive to conserve.)

## Domestic hot-water systems

- Choose shower heads that provide and maintain user comfort and energy savings. They should not have removable flow-restricting inserts to meet flow limitation requirements.
- Consider point-of-use water heaters where their use will reduce energy consumption and annual energy cost.

- Consider using storage to facilitate heat recovery when the heat to be recovered is out of phase with the demand for hot water or when energy use for water heating can be shifted to take advantage of off-peak rates.

## Controls

Well-designed digital control provides information to managers and operators as well as to the data processor that serves as the intelligent controller. Include the energy-saving concepts discussed previously throughout the operating sequences and control logic. However, energy conservation should not be sought at the expense of inadequate performance; in a well-designed system, these two parameters are compatible.

**Keywords**: Energy use, Energy-efficient design, Energy conservation

**Source from**:
ASHRAE. 2009 ASHRAE Handbook. Fundamentals-Chapter 35. Atlanta, GA: ASHRAE, 2009.

## Part 3　Extension

1. Building energy use(建筑能耗)

建筑能耗有两种定义方法:广义建筑能耗是指从建筑材料制造、建筑施工,一直到建筑使用的全过程能耗;狭义建筑能耗或建筑使用能耗则是指维持建筑功能和建筑物在运行过程中所消耗的能量,包括照明、采暖、空调、电梯、热水供应、烹调、家用电器以及办公设备等的能耗。除非特别指明,现在一般提及的"建筑能耗"都是指使用能耗。

2. HVAC system design(暖通空调系统设计)

暖通空调系统的选择及其设计,将直接影响其最终的能耗。暖通空调系统的节能设计应当做到:①详细进行系统的冷热负荷计算,力求与实际需求相符,避免最终的设备选择超过实际需求,否则既增加了投资,也不节能;②选择高效的冷热源设备;③减少输送系统的动力能耗;④选择高效的空调机组及末端设备;⑤合理调节新风比;⑥采用热回收与热交换设备,有效利用能量。

## Part 4　Words and Expressions

| insulation | *n.* 绝缘;隔声;隔离,孤立;绝缘或隔热的材料 |
| --- | --- |
| nonrenewable resource | 不可再生资源 |
| replenishment | *n.* 补给,补充 |
| energy use | 能源消费,能源利用 |
| energy-efficient design | 节能设计 |

续表

| heating and cooling loads | 供热和供冷负荷 |
| --- | --- |
| sensible cooling | 显冷却；显冷；等湿冷却 |
| thermal power | 火力；热动力,热功率,热能 |
| hydronic | *adj.* 液体循环加热(或冷却)的 |
| water chiller | 冷水机组 |
| chilled-water system | 冷冻水系统 |
| unitary air-conditioning system | 整体式空调系统 |
| orientation | *n.* 方向,定位,定向,排列方向；任职培训 |
| window/wall radio | 窗墙比 |
| carbon offsets | 碳补偿；碳抵消；购买碳抵消额度；碳平衡 |
| design load | 设计负荷 |
| part-load | 部分负荷 |
| peak-load | 峰值负荷 |
| diversified energy | 多样化能源 |
| thermal conductivity | 导热系数 |
| phase-change | 相变 |
| thermal bridge | 热桥 |
| building envelope | 围护结构 |
| comprehensive design | 综合设计；整体设计 |
| opaque | *adj.* 不透明的；无光泽的,晦暗的；不传导性的；含糊的,迟钝的 *n.* 不透明,晦暗；[建]遮檐；遮光涂料 |
| dew point | 露点 |
| solar radiation | 太阳辐射 |
| daylighting | 采光 |
| shading coefficient | 遮阳系数 |
| radiation shield | 辐射屏蔽 |
| heat-generating equipment | 发热设备 |
| heat recovery | 热回收 |
| life-cycle | 生命周期 |
| two-phase | 两相 |
| single-phase | 单相 |
| air duct | 风道 |
| critical pressure | 临界压力 |

## Part 5　Reading

## Energy Efficiency for HVAC&R Systems

### 1. *Energy use and energy efficiency for HVAC&R systems*

Energy use or energy consumption indicates the amount of energy used or consumed. Energy efficiency indicates how efficiently energy is used. An energy-efficient HVAC&R system maintains a comfortable indoor environment with an acceptable indoor air quality (IAQ) and consumes optimum energy resources. According to the ASHRAE definition, energy conservation indicates that energy is used efficiently. Energy management is the effort and measures taken to ensure that energy is used efficiently, and the unit energy rates (electric and gas rates) are reasonably low for the sake of reduction of energy cost. Energy management of HVAC&R systems consists of two areas: energy efficiency (the reduction of energy use or energy conservation) and the reduction of the unit energy rate.

The estimate of annual U. S. energy use of the HVAC&R systems in 1992 was approximately one sixth of the total national energy use. In addition, energy use for HVAC&R is closely related to the release of $CO_2$ to the outdoor atmosphere which causes the global warming effect. Energy efficiency is a challenge to every one of us in the HVAC&R industry now and for many years to come in the future.

According to EIA Commercial Building Characteristics 1992, energy sources used for the heating of the 67.8 billion $ft^2$ of commercial buildings in the United States in 1992 were electricity, 22.8 percent; natural gas, 51.8 percent; fuel oil, 6.6 percent; and district heating, 7.3 percent. Energy sources used for cooling U. S. commercial buildings in 1992 were electricity, 80.4 percent; natural gas, 2.8 percent; and district cooling, 3.0 percent. Electricity and natural gas are the two primary energy sources for HVAC&R systems.

### 2. *Energy efficiency during design, construction, commissioning, and operation*

Energy efficiency must be achieved in every stage of HVAC&R system construction and operation: design, construction, commissioning, and daily operation. A well-designed and effectively functioning energy management direct digital control (DDC) system is necessary for an energy-efficient HVAC&R system.

- Design. Various alternatives should be compared and analyzed in terms of either payback years or life-cycle cost to determine which is most energy-efficient and cost-effective.
- Construction and installation. Roof and external walls should be well insulated. The efficiency of installed equipment should not be less than the minimum efficiencies specified by ASHRAE/IESNA Standard 90.1—1999. The amount of air leakages from the ducts, and the duct and pipe insulation all affect the energy use.

• Commissioning. The capacity of equipment, the air and water balance, and the coordination between various components and control systems should be carefully measured, adjusted, and commissioned. A poorly commissioned HVAC&R system will never function efficiently as specified.

• Operation. The energy use for chillers, compressors, fans, pumps, boilers, and furnaces should be monitored, periodically checked, investigated, reduced, and improved.

For HVAC&R, to reduce the emissions of $CO_2$ to the atmosphere by means of increasing energy efficiency in operation is the primary action to mitigate the global warming effect.

## 3. *Reduction of unit energy rates*

Facility owners, facility managers, or the tenants as well as the designers and operators of the HVAC&R systems all are concerned about the unit energy rate $E_r$, as it affects the energy cost and the operating cost even if there is no saving in HVAC&R system energy use. For electricity, energy cost $C_e$ is the product of the price of unit energy rate $E_r$ and energy use, in kW·h or therms. The lower $E_r$, the smaller $C_e$. The reduction of unit energy rate $E_r$ is closely related to the following:

• The ratio of unit rate of kW·h to therms which affects the designer in selecting electric cooling or gas cooling.

• Use of thermal storage to provide off-peak conditioning to reduce $E_r$.

• Choice of most favorable and optimum $E_r$ rate from the electricity deregulation or gas deregulation Energy cost is the dominant factor that affects the operating cost of an HVAC&R system.

## 4. *Energy conservation measures*

In principle, adopting an integrated design approach by considering the building as a whole, minimizing heating and cooling loads, selecting high-efficiency equipment sized as closely to the design load as possible, emphasizing commissioning, and optimizing operation promote energy efficiency (energy conservation) in buildings. The following are possible energy conservation measures (ECMs) for the HVAC&R system in buildings:

• Turn off electric lights, office appliances, and other equipment when they are not needed.

• Shut down AHUs, PUs, fan coils, VAV boxes, chillers, fans, and pumps when the space they serve is not occupied or when they are not needed, except during warm-up, cool-down, or the purging period prior to the morning occupied period.

• The time chosen to start or to stop the AHUs, PUs, chillers, and exhaust fans daily should be optimum.

• The set point of space temperature, relative humidity, cleanliness criteria, and indoor-outdoor pressure differential should be optimum. Using different space temperature set points for summer and winter seasons for comfort air conditioning is often energy-efficient.

• The discharge air temperature from the AHU or PU and the temperature of water leaving the chiller or boiler should be reset for part-load operation based on the space temperature, system load, or outdoor temperature.

• Reduce air leakages from ducts, dampers, equipment, and HVAC&R system components.

Use weather stripping to seal windows and revolving exterior doors to reduce infiltration. Ducts and pipes should be well insulated. Carefully design the layout of ducts and pipes to minimize their length, the number of duct and pipe fittings, as well as their pressure losses.

- Use more energy-efficient cooling methods such as free cooling, evaporative cooling, or groundwater cooling instead of refrigeration. Replace refrigeration with an air economizer, water economizer, evaporative cooler, and even desiccant dehumidification if doing so provides the same cooling results, is more energy-efficient and is cost-effective. An evaporative condenser is often more energy-efficient than a water-cooled or air-cooled condenser in many U.S. locations.

- Use heat recovery systems, waste heat from gas-cooling engines, or heat pumps to provide winter space heating when they are applicable and cost-effective.

- Use variable-air-volume systems instead of constant-volume systems, and variable-flow building loop water systems instead of a constant-flow one if the airflow or the water flow should be reduced during part-load operations. Large variable-speed aerofoil centrifugal fans are energy efficient and often cost-effective in VAV systems.

- Use energy-efficient chillers, boilers, AHUs, PUs, and motors. Todesco (1996) recommended that chillers have an energy use index of 0.50 to 0.55 kW/ton, condensing boilers of 90 percent efficiency, and large supply and return fans of 65 percent total efficiency and higher.

- Adopt a direct digital control (DDC) energy management and control system for large and medium-size HVAC&R new and retrofit projects.

- Adopt double-pane windows with low-emission coatings. Todesco (1996) recommended a $U$ value of 0.25 Btu/h·ft$^2$·°F [1.4 W/(m$^2$·°C)] for exterior windows, also lower U values for external wall and roofs. Reduce the heat gain in the summer and heat loss in the winter in locations where the outdoor temperature is high in the summer and low in the winter.

- Use daylighting and controls for the perimeter zone. Also use energy-efficient fluorescent lamp and electronic ballasts with a goal to achieve connected lighting loads of 0.75 W/ft$^2$ (8 W/m$^2$) or better.

## 5. Case study: energy conservation measures for office

Parker et al. (1997) introduced an energy-efficient office building design for the Florida Solar Energy Center (FSEC) in the hot and humid climate of Cocoa, Florida. This office building consists of offices, a visitors' center, and laboratories and has a floor area of 41,000 ft$^2$ (3,809 m$^2$). Laboratories were not included in this energy analysis. Eighteen energy conservation measures (ECMs) were considered during the energy analysis. Some of the ECMs are not effective for the hot, humid climate in Florida; e.g., the air economizer cycle showed little savings, and higher than code-mandated levels of wall and roof insulation resulted in little advantage. Only the following ten ECMs provided cost-effective energy savings.

- A high-efficiency lighting system of 0.9 W/ft$^2$ (9.7 W/m$^2$) with T-8 fluorescent lamps in a reflective troffer and with electronic ballasts saves energy use.

- Windows comprised of glazing units with high visible transmittance of 0.56 for daylighting and a low shading coeffficient of 0.33 reduce unwanted solar heat gain. Also a low $U$ value of 0.31 Btu/h·ft²·°F [1.76 W/(m²·℃)] reduces conductive heat gain.

- Daylighting perimeter illumination coupled with solar control system on the south facade enables effective use of the perimeter office illumination. Dimming electronic ballasts are controlled by photometric sensors to adjust ballast output combined with the available daylight to maintain a constant desktop illumination level.
- Two high-efficiency screw chillers provide an IPLV of 0.65 kW/ton ($COP = 5.41$) and of 0.60 kW/ton ($COP = 5.86$) at 50 percent part load.
- Energy star personal computers, printers, and copiers are used to save energy.
- A reflective white single-ply roof membrane is chosen instead of a gray or black one, to reduce the solar irradiated heat gains through the roof.
- Use of occupancy sensors shuts off the VAV terminals when a room is not occupied.
- Using a DDC EMCS permits an increase in the cooling set point from 75 to 76 °F (23.9 to 24.4 ℃) because of a finer control tolerance as well as provides optimal start and stop capability.
- Using variable-speed fans and variable-speed building pumps reduces the fan speed or pump speed during part load to save fan and pump energy use. Analysis showed that they are cost-effective.
- $CO_2$-based demand ventilation control is adopted for intermittently occupied spaces and zones.

In Florida, the average energy use of 160 state-owned office buildings in 1991 was 67 MBtu/h·ft²·yr (211 kW·h/m²·yr). With these ten ECMs, the predicted energy use for the office building of FSEC will drop to 27 MBtu/h·ft²·yr (85 kW·h/m²·yr).

**Source from:**
Shan K. Wang. Handbook of Air Conditioning and Refrigeration-Chapter 25. New York: McGraw-Hill, 2000.

## Part 6 Exercises and Practices

### 1. *Discussion*

(1) Discuss the main building energy use elements.
(2) How to consider the energy efficiency plan in HVAC system design?

### 2. *Translation*

(1) Extensive study since the mid-1970s has shown that building energy use can be significantly reduced by applying the fundamental principles discussed in the following sections.
(2) The basic approach to energy-efficient design is reducing loads (power), improving transport systems, and providing efficient components and "intelligent" controls.
(3) Control thermal conductivity by using insulation (including movable insulation), thermal mass, and/or phase-change thermal storage at levels that minimize net heating and cooling loads on

a time-integrated (annual) basis.

(4) Consider separate HVAC systems to serve areas expected to operate on widely differing operating schedules or design conditions.

(5) Select the most efficient (or highest-COP) equipment practical at both design and reduced capacity (part-load) operating conditions.

## 3. Writing

Please list the basic approaches to energy efficiency design in 200~300 words.

# Lesson 19
## Big Data in HVAC

### Part 1　Preliminary

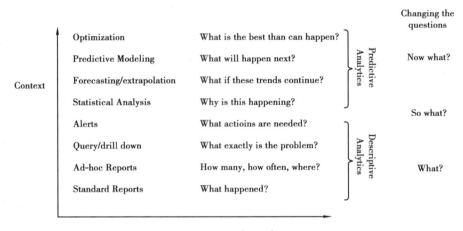

Fig. 19.1　How Big Data changes the questions

Here are some examples of how we can use Big Data in HVAC systems. On one hand, it helps to create output in a progressive manner. The left part of the figure focuses on data, while the right part tries to explain the data and find out the root of problems. On the other hand, we can use that output to answer more analytical questions, which brings increasing value. The bottom part of the figure indicates the conventional way that only analyzes the current state, while the upper part concentrates on future trend. In summary, Big Data provides the possibilities of moving from data to analytics, from descriptive analytics to predictive analytics.

### Part 2　Text

大数据是指新型处理模式下具备更强的决策力、洞察发现力和流程优化能力的大容量、高增速和多样化的信息资产。近年来，智能设备和数字化的发展增大了数据的数量、种类和复杂性，因此对数据的筛选和有效利用十分重要。自动控制技术的飞速发展，计算机处理、存储成本的降低和能力的提升，使得空调系统、企业和用户为暖通空调领域的大数据提供了广泛的数据来源和用途。大数据分析能够预测空调能耗，诊断并指导运行调控，实现空调系统整体及其构件的优化。可用于改善产品开发，提高产品质量，减少产品研发周期，提升企业竞争力。还可以根据用户使用空调的规律，分析用户需求，提供室内环境的个性化调节及互联网、物联网的智能控制服务。

## 1. Big Data

**Big Data**[①] refers to high-volume, high-velocity and/or high-variety information assets that require new forms of processing to enable enhanced decision making, insight discovery and process optimization. How big is it? Big Data is measured in terabytes(1,000 gigabytes), petabytes (1 million gigabytes) and zettabytes (1 trillion gigabytes).

The world is full of examples of Big Data. Walmart in the United States handles about a million credit card transactions every hour. Across the U.S., there are more than 40 billion credit card transactions every year. Using more than 60 years of historic data and models involving 82 billion calculations, one startup company is now able to accurately predict weather conditions 40 days in advance. IBM interprets 350 billion annual meter readings to better predict power consumption.

Globally in 2012, 2.7 zettabytes of digital content was created. The advent of smart equipment and greater digitization significantly has increased the volume of data points available from the field. Besides the volume, complexity of data and establishing interrelationships is a huge challenge. Globally, $28 billion of IT spending in 2012 was on Big Data technologies, and is expected to increase to $34 billion in 2013.

Over the last few years, it is not just the volume of data that has increased exponentially; it is also the variety — we capture more variables today on top of capturing more samples of each variable.

Taking the example of buildings — the traditional approach to energy-baseline definition would have been to correlate whole building consumption with outside air temperature, maybe also occupancy and level of operations. However, with advanced metering and building automation systems, we now can analyze data by individual service type, specific building location and even specific time of the day. We therefore have more variables to take into account (and of course the large volume of data) for the same analysis—which is great as it allows for more targeted analysis, but which also means that we need to identify the most important variables for analysis, as otherwise we could easily be lured into misleading analytics.

Big Data always has existed. Today, technology allows us to capture and use Big Data to connect various elements of the ecosystem — facility data, performance and operational data of the systems installed in these connected facilities, maintenance histories and manufacturers' data. Advancements in technology are making it easier to connect, capture and use Big Data. Technology finally has evolved to allow us to digitize building subsystems and make them transportable virtually as Big Data.

Therefore, the challenge of **Big Data analytics**[②] really starts with big data management—and then the analytics. Big Data analytics must in turn lead to "big insights" and "big actions"—the challenge today is in completing the "Big Data and Analytics Value Chain".

---

①② 见 Part 3 Extension.

## 2. Big Data in HVAC: the case of chillers

In a building, the heating ventilation and air conditioning (HVAC) system typically consumes about 60 percent of the building's total energy requirement. Of this, chiller plants consume 35 percent — translating to 20 percent, or one-fifth of the total building energy use.

So, chiller plants must be one of the first logical starting points for any building energy-efficiency initiative. On the average, about 30 **operating parameters**[③] of a chiller can be monitored. If these parameters are captured and recorded every 15 minutes, it translates into more than 1 million records a year for a single chiller. For a medium-sized building, normally about 600 HVAC equipment and system data points are captured using **building automation systems (BAS)**[④]. If the information is captured every 15 minutes and stored for each building, there will be over 21 million records every year. For 5,000 similar buildings, over 105 billion records will be captured annually, translating into over 4.2 terabytes of data storage.

Building owners and operators can leverage Big Data and Analytics in many ways to create tangible value:

**Advanced analytics.** It can help with better understanding of building and equipment performance. It allows historical trending, pattern recognition and correlation between cause and effect of issues and events occurring in the various building and HVAC subsystems.

**Intelligent insights.** It enables benchmarking of a building's HVAC system performance against industry standards or benchmarks. Owners and operators can cross check what their energy usage is in HVAC systems and how they stack up against peers.

**Preventive maintenance.** Through proper analytics on past performance data and issue trends, future potential maintenance issues can be identified through simulation and predictive technologies. Such actions will help extend equipment life, reduce operating costs and minimize disruption.

**Informed decisions.** Leveraging Big Data and Analytics, building managers also can model their future energy requirements and simulate their future operating budgets.

**Value asset.** Monetize raw data for parties interested in sustainability such as educational institutes, research bodies and policy groups.

**Connected communities.** At a fundamental level, virtualization of building subsystems allows harnessing dispersed experts by creation of a connected community of advisors to enhance performance of buildings. "Bringing the building to subject matter experts" is now possible through organized Big Data and the Internet.

Big Data and Analytics brings much additional value for HVAC systems manufacturers and service providers:

- Improve future design by understanding how their equipment and systems are used by customers, facilitating the alignment between product development and customer needs;
- Anticipate future repair and replacement needs, thus improving service quality and plan-

---

③④ 见 Part 3 Extension.

ning;
  • Increase service productivity with more accurate targeting of current and future issues identified by analytics on Big Data;
  • Differentiate their relationship and service offerings from organizations that are not leveraging Big Data and Analytics, thereby creating new economic models.

Big Data and Analytics helps us understand what is going on with buildings and the HVAC systems in them, the implications of that and what kind of actions are recommended to improve building performance. This is achieved as we progress on the dimensions of understanding the data and putting it in the context in which the data is created. As we get better in moving from data to analytics, we move from more descriptive analytics to predictive analytics.

## 3. *The urgency around Big Data and Analytics*

Today, it has become imperative for building owners/operators and OEM/service providers to start making use of Big Data and Analytics. There are several key drivers for this:

Technology in "digitization of buildings" is now a reality. Customers and building operators should be able to demand to be supported and served by the subject matter experts and best advisors from OEMs and service providers.

Reliance on a building and HVAC system performance is becoming critical to a company's success as it helps with occupant health, comfort, productivity and compliance.

In several industries, including pharmaceutical, food processing and scientific research, temperature and environment control of the facilities is critical for proper conduct of business. This makes the HVAC systems even more vital for such applications beyond health and comfort as the revenue of such industries is fully dependent on the facility environment.

Many countries have statutory requirements around energy usage. To comply with most of these statutes, building owners must submit fairly detailed proof of efficiency and performance of systems, and often provide aggregated results from those.

Most corporations are becoming more involved in their sustainability initiatives with the commitment coming straight from senior management. Big Data and Analytics creates technology and information platforms for the facility managers and sustainability leaders in organizations to drive sustainability initiatives.

Big Data and Analytics provides some alternative options to capture and institutionalize expertise on HVAC systems in a world competing for top talent.

Organized big data from buildings now need to be managed also by IT teams along with other business groups such as finance, HR, legal and knowledge management. There should be similar considerations on security and access. This is the convergence of building systems data with information technology.

## 4. *How to handle Big Data and Analytics*

Addressing opportunities and challenges around Big Data and Analytics in the space of building and HVAC systems requires a systematic approach. Some key steps involved in this process are:

**Capture** — connect and collect information from different equipment and building sub-systems;

**Curate** — select and organize collected information;

**Manage** — store and correlate information to derive knowledge and wisdom;

**Process** — analyze and present information in an actionable format with economic impact indicators.

The HVAC and buildings industry is evolving in interesting ways in response to the opportunities around Big Data and Analytics. Almost all the large equipment/system manufactures and service companies such as Johnson Controls, Honeywell, Carrier, Trane, Schneider, Siemens, Daikin and Lennox are investing aggressively in this space to become the market leader. This space is also seeing interest from mechanical contractors and facilities management companies. Large government and private organizations have undertaken their own initiatives around Big Data and Analytics to further their sustainability agenda or optimize their energy and operational budgets. Most large IT companies such as IBM, Wipro and SAS are also moving fast to build capabilities and solutions in this space. They are incorporating learning from different industry domains which also create and use Big Data to apply to the HVAC and buildings industry. Accessibility of the enabling technology is leveling the playing field.

## 5. *Challenges in developing strategies*

Big Data and Analytics is quite matured in certain industry domains such as banking, financial services, retail, defense and security. However, in the HVAC and buildings industry, it is still in its early days but evolving rapidly. Companies invested in Big Data and Analytics are addressing several challenges posed by the uniqueness of this discipline:

• Organization of building data to enable analytics and advisory services is one of the biggest challenges to overcome in this industry.

• Buildings and HVAC systems within them are dispersed over geographic expanses. Remotely connecting such sites to collect data in an inexpensive manner with an industrialized approach is a problem which many companies are trying to address.

• Companies are also exploring effective solutions to store and manage the huge volume of data once collected.

• Variation of scope, design and configuration leads to differences in HVAC systems that increase the challenges of data normalization.

• After one is able to capture and curate the data, the next challenge of developing flexible platforms is to create visualization reports and dashboards on the fly to meet individual needs.

• Deploy robust analytical platforms for predictive modeling, fault detection and diagnostics, and create economic value from Big Data and Analytics.

The core technology for HVAC has been stable over many years, but Big Data and Analytics looks set to bring the HVAC industry through a paradigm shift, especially with respect to how buildings are viewed, operated, managed and serviced. Big Data and Analytics is here and now. People already have started investigating the possibilities and investing in future. This is an area which needs to be understood well, but acted upon quickly.

**Keywords**: Big Data and Analytics, HVAC system, chiller, energy efficiency

**Source from**:
Sudhi S, Snehil T and Swarup B. *How Big Data and analytics change the game for HVAC systems*. GreenBiz, 2013.

## Part 3　Extension

1. Big Data(大数据)

大数据是指无法在一定时间内用常规软件工具对其内容进行抓取、管理和处理的数据集合,具有数据体量巨大、类型多样、处理速度快、价值密度低的四个基本特征。大数据技术是指从各种各样类型的数据中,快速获得有价值信息的能力。适用于大数据的技术,包括大规模并行处理数据库、数据挖掘电网、分布式文件系统、分布式数据库、云计算平台、互联网和可扩展的存储系统。

2. Big Data Analytics(大数据分析)

大数据分析主要包括六个方面:可以直观展示数据的可视化分析,深入数据内部的数据挖掘算法,让分析员做出预测分析判断的预测性分析能力,从"文档"中智能提取信息的语义引擎和对数据进行整合的数据仓库。

3. Operating Parameters(运行参数)

冷水机组的运行参数包括冷冻水的供回水温度、冷冻水流量等,它们与机组的运行效果、能耗使用情况紧密联系。对大量的数据进行处理分析,可以判断机组是否能提供足够冷量、是否节能,从而为冷水机组的运行模式提供可靠依据。

4. Building Automation System (BAS)(楼宇自动化系统)

将建筑物或建筑群内的电力、照明、空调、给排水、消防、运输、保安、车库管理设备或系统,以集中监视、控制和管理为目的而构成的综合系统。系统通过对建筑(群)的各种设备实施综合自动化监控与管理,为业主和用户提供安全、舒适、便捷高效的工作与生活环境,并使整个系统和其中的各种设备处在最佳的工作状态,从而保证系统运行的经济性和管理的现代化、信息化和智能化。

## Part 4　Words and Expressions

| | |
|---|---|
| transaction | *n.* 交易;事务;办理 |
| Interrelationship | *n.* 相互关系;干扰 |
| exponentially | *adv.* 以指数方式;按指数规律地 |
| correlate | *vi.* 关联 |
| | *vt.* 使有相互关系;互相有关系 |
| | *n.* 相关物;相关联的人 |
| metering | *n.* 计量;测量;测光模式 |

续表

| evolve | v. 发展；进化；逐步形成 |
|---|---|
| initiative | n. 新方案；倡议；主动权；首创精神 |
| | adj. 主动的，自发的，起始的 |
| leverage | v. 利用；举债经营 |
| | n. 手段，影响力；杠杆作用；杠杆效率 |
| tangible | adj. 有形的；切实的；可触摸的 |
| | n. 有形资产 |
| disruption | n. 破坏；毁坏；分裂；瓦解 |
| monetize | v. 赚钱；盈利；定为货币；铸造成货币 |
| harness | vt. 利用；治理；套；驾驭 |
| | n. 马具；甲胄；降落伞背带；日常工作 |
| dispersed | adj. 散布的；被分散的；被驱散的 |
| alignment | n. 队列，成直线；校准；联盟 |
| imperative | adj. 必要的，不可避免的；紧急的；专横的，势在必行的；[语]祈使的 |
| | n. 必要的事；命令；需要；规则；[语]祈使语气 |
| pharmaceutical | n. 制药；药物 |
| | adj. 制药(学)的 |
| statutory | adj. 法定的，法令的；可依法惩处的 |
| comply | vi. 遵守；遵从；顺从；答应 |
| Aggregated | adj. 合计的；聚合的 |
| Convergence | n. [数]收敛；会聚；集合 |
| Curate | n. 助理牧师；副牧师 |
| Contractor | n. 承包商；立契约者 |
| Domain | n. 领域；域名；产业；地产 |
| Dashboard | n. 仪表盘；汽车等的仪表板 |
| Deploy | vi. 部署；展开 |
| | vt. 配置；展开；使疏开 |
| Robust | adj. 强健的；健康的；粗野的；粗鲁的 |
| Paradigm | n. 范例；范式；词形变化表 |
| take into account | 考虑到；把……计算在内 |
| cross check | 交叉校验；交叉检查；交互核对 |
| stack up against | 与……相比；争胜负；较量 |
| raw data | 原始数据，原始资料 |
| leveling the playing field | 创造公平的竞争环境 |
| OEM (original equipment manufacturer) | 原始设备制造商 |

## Part 5　Reading

## Design and Implementation of Intelligent HVAC System Based on IoT and Bigdata Platform

### 1. Introduction

Through the improvement of high-tech industries and enhancement of wired and wireless networks, home appliances once provided simple and fundamental services for consumers. Through Internet linkage, monitoring and remote control services are now Internet of Things (IoT) device that are made available to consumers. IoT refers to technologies such as low power wireless communication modules likeWIFI, Bluetooth Low Energy (BLE), ZigBee, and technologies that collect and utilize data such as room temperature, humidity, intensity of illumination to provide user customized-services.

Recently various devices that utilize IoT technologies havebeen released, and people install HVAC based IoT such as Thermostats to improve home environment. There are studies related to the wireless communication modules and HVAC control scheduling methods for reducing building HVAC energy consumption, and ongoing studies on how to optimize energy control through HVAC simulation analysis.

Current research is efficient from an energy consumption reduction point of view. However, HVAC has limitations and problems from a user convenience point of view as well as provision for various services.

First, in order to install a HVAC system, the remote controldevice needs to be installed in the living area or entrance of the household, and must be controlled manually. Due to suchmethods, users face the inconveniency of having to move the control device to set up the value for the current environmental setting. Second, the existing HVAC system does not provide user custom services, but only provides area or group specific services. Finally, while the HVAC services are applied in large buildings, in smaller buildings and households must purchase HVAC devices, and additional expenses for construction may arise. For these reasons, the user is in need of a HVAC system that can be installed andoperated through cheap costs.

In order to solve such problems, this paper suggests to constitute an automatic control system through current air-conditioners, fans, ventilators, and heating devices withouthaving to purchase a new HVAC system. Also, this papersuggests IoT and Bigdata based intelligent air-conditioning system that will enable communication between smartphones and sensor boxes and provide user customized environmental services.

### 2. System architecture

Fig. 19.2 represents the Intelligent HVAC System based onIoT and Bigdata platform (IHSIB) architecture suggested inthis paper. IHSIB is constructed with a sensor box, server, Smart plugs (Wi-Fi based plugs) and remote controls. Among these, the Wi-Fi based plug and remote control was developed to control air conditioners, heating devices, and ventilators. The sensor box is set up in each individual space, and when the user enters the particular space, the sensor box recognizes the user and provides user-customized services. Through this paper, it explains that the user customized control service and real time environment monitoring service developed through IHSIB.

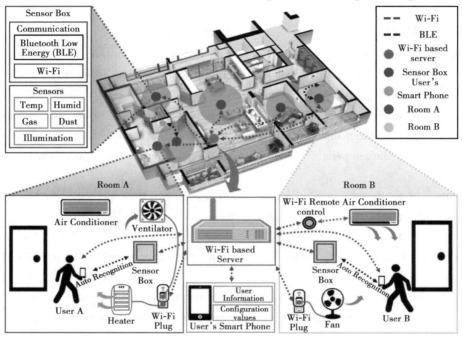

Fig. 19. 2　Overview of the proposed system architecture

Fig. 19. 3 represents IHSIB system's control sequence diagram. When the user enters a specific space such as a room, the IHSIB sensor box utilizes the user's smartphone BLE signal to recognize that the user has entered the specific space. The smartphone uses an application to automatically send the usersID value to the sensor box, and the sensor box sends the ID value and environmental information collected in real time to the server through Wi-Fi communication. Through the order delivered by the sensor box, the server utilizes the Wi-Fibased plug and remote to control HVAC devices such as air conditioners, heaters, and ventilators in order to meet the environment value the user selected. When the user leaves theroom or exits, the IHSIB Sensor Box recognizes the absence and sends message to server to turn off the HVAC device. Also, when the IHSIB Sensor Box registers that there are various users in the same space, the device refers to the environmental value saved on the server administrators application to control the HVAC.

Fig. 19. 4 explains the algorithm about IHSIB sensor box. When the user leaves the room for external business, HVAC devices stop operating, but the sensor box continues to collect information such as room temperature, humidity, intensity of illumination, dust level, and gas. When the sensor box detects gas during operation, the sensor box alerts the server of the situation and the server informs the user and organization in the vicinity with an emergency message in order to prevent user property loss.

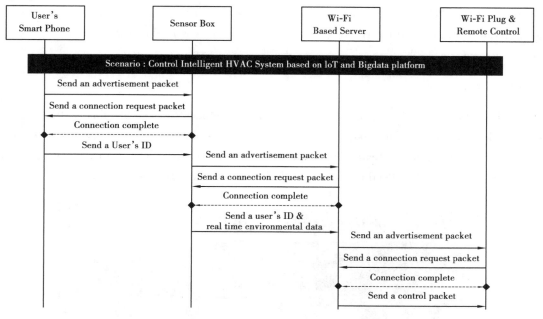

Fig. 19.3　Sequence diagram about IHSIB

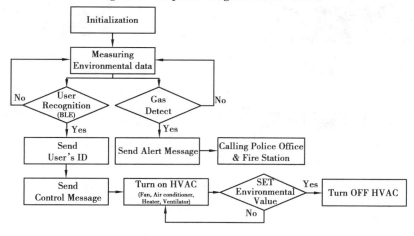

Fig. 19.4　The algorithm about IHSIB sensor box

## 3. Implementation

Fig. 19.5 represents the suggested IHSIB prototype. This system uses hardware and software based on open source IoT platforms, and the IHSIB sensor box's main board is constructed around RaspberryPi3 and the sensor. The corresponding board was used to take in real time sensors. Also, the sensors that measure room temperature, humidity, lighting, dust levels, and gas are utilized for both real time environment measurements.

Fig. 19.6 represents IHSIB's smartphone application and server control application. First, Fig.

19.6 (a) shows thesmartphone application. Through this application, user information and desired environmental value is saved and managed. Fig. 19.6 (b) shows the IHSIB system administrator server application. The environmental information of the space where the sensor box is set up can be monitored in real time, and the HVAC can set up the environmental value when there are various users in the same space.

## 4. Conclusion

In this paper, it suggests that through IoT based intelligent air-conditioning systems, a user customized control service and real time monitoring service is able to provide to users, thus reducing inconveniences. For future research, we hope to conduct research on how to sophisticate user customized services through collecting and analyzing information through Bigdata analysis.

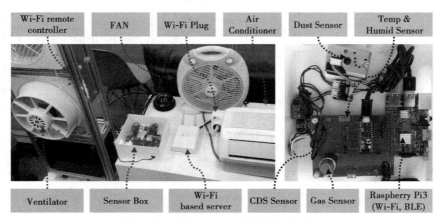

**Fig. 19.5 Prototype of IHSIB device**

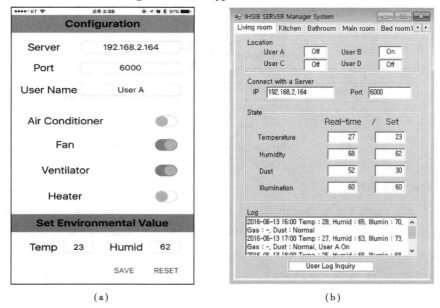

**Fig. 19.6 Applications of Intelligent HVAC System**

**Source from:**
Lee, Tacklim, et al. *Design and implementation of intelligent HVAC system based on IoT and Bigdata platform.* IEEE International Conference on Consumer Electronics, 2017.

## Part 6    Exercise and Practice

### 1. Discussion

(1) How can records from chillers be used to create tangible value for building owners and operators?

(2) What are the challenges companies invested in Big Data and Analytics are facing?

### 2. Translation

(1) Using more than 60 years of historic data and models involving 82 billion calculations, one startup company is now able to accurately predict weather conditions 40 days in advance.

(2) The advent of smart equipment and greater digitization significantly has increased the volume of data points available from the field.

(3) Technology finally has evolved to allow us to digitize building subsystems and make them transportable virtually as Big Data.

(4) So, chiller plants must be one of the first logical starting points for any building energy-efficiency initiative.

(5) Big Data and Analytics helps us understand what is going on with buildings and the HVAC systems in them, the implications of that and what kind of actions are recommended to improve building performance.

### 3. Writing

Check out relevant materials to introduce the application of Big Data in other parts of the HVAC system besides chillers.

# Lesson 20
# Engineering Ethics and Professional Morality

## Part 1  Preliminary

**Engineering**① is a **profession**②, similar to law, medicine, dentistry, and pharmacy, which is highly educated and specialized expertise. Technology almost always raises new moral and social issue—or, most commonly, old issues in new ways. For example, the occupant of the Tesla driverless car was killed when a tractor-trailer made a left turn in front of the car. The car went under the truck's trailer without applying the brakes, evidently because neither the autopilot nor the drive noticed the white side of the trailer against a brightly lit sky. Investigation revealed that the driver did not operate the Tesla according to instructions, and that Tesla did not deploy a system capable of identifying situations in which the driver was not "prepared to take over at any time." Therefore, where should moral responsibility and legal liability lie in this case.

Many questions have appeared in other forms and other contexts before. Moral issues also arise in thinking about the duties of engineers in such areas as the relationship of technology to the environment and handling risk property. The issues are important to engineers not simply because engineers have usually created the technologies involved, but because engineers are professionals, and the concept of professionalism has a strong moral component.

## Part 2  Text

工程伦理是工程技术人员的职业道德。由于工程具有区别于其他职业的重要特征，所以开展工程伦理教育有着重大意义。尤其，现代化工程的出现，其专业间合作要求高，社会效应和社会影响巨大，对环境和人类未来的影响十分重大。随着中国经济快速发展和许多重大工程项目的实施，越来越多的工程伦理问题凸显在我们面前，如工程质量、公共安全、工程与环境、工程与生态、工程师的科学态度和职业精神等问题，已经成为社会重点关注的问题。因此，在推进工程质量化和社会和谐化的背景下，职业伦理，特别是工程伦理备受关注。

### 1. *Engineering and profession*

**About profession**

First, there is the Sociological Account, which holds that there are characteristics especially associated with professionalism. Second, professionals have an implicit agreement with the public on the Social Contract Account. Professionals agree to attain a high degree of professional expertise, to

---

①② 见 Part 3 Extension.

provide competent service to the public, and to regulate their conduct by ethical standard. And the public agrees to allow professionals to enjoy above-average wages, to have social recognition and prestige, and to have a considerable degree of freedom to regulate themselves. A third account of professionalism is offered by philosopher Michael Davis, who defines a profession in the following way: "A profession is a number of individuals in the same occupation voluntarily organized to earn a living by openly serving a moral ideal in a morally permissible way beyond what law, market, morality, and public opinion would otherwise require.[1]"

**Engineering is a profession**

Engineering is clearly a profession by all three accounts. When considering the Sociological Account, to be an engineer requires high level of training at the college or university level. Engineering is vitally important to the public. One cannot imagine our society without highways, airplanes and many other technological artifacts designed by engineers. Engineers have considerable control over the curriculum in engineering schools and the standards for admission to the profession.

To continue, while engineers who work in business and public organizations may not be as autonomous as lawyer or doctors who have their own practice, they probably have more autonomy than most nonprofessionals, if only because non-engineers do not have enough technical knowledge to give more than general direction to engineers. Like other professionals, engineers have ethical codes that are supposed to regulate their conduct for the public good. Someone may claim that professional codes are mere window dressing, designed to disguise the fact that professionals are primarily out to promote their own economic self-interest. Even there is some truth to the claim, it should be believed that ethical considerations are taken very seriously by most engineers and other professionals.

Engineers in general have a high level of professional expertise and render competent service. Engineers also have ethical codes, but the loss of PE registration as a penalty for unethical conduct does not prohibit an engineer from professional practice, as in most other professions, since engineers are not required to be licensed to practice. So perhaps it can be said that the engineering profession does not have the same ability to enforce ethical sanctions as some other profession. Nevertheless, a server ethical violation can tarnish the reputation of an engineer and possibly subject the engineer to legal penalties.

## 2. Ethic and morality

**Virtue ethic**

Virtue ethic, perhaps the oldest tradition of ethical thought, has become increasingly important among contemporary ethicists. The fundamental principle of virtue ethics is "Act in the way the good or virtuous person would act in the circumstances."

A virtue is usually described as a "dispositional trait", that is, a character trait that disposes or inclines a person to do the right thing. A virtue can be described as both deep and wide. It is deep in the sense that a virtue is a firmly entrenched habit that leads a person to consistently act in a certain way and to which he is strongly committed. It is wide in that it manifests itself in a variety of ways.

Virtue ethicists such as Aristotle have found it useful to think of virtues as occupying a middle position between vices. We can think of courage as a middle ground between the vice of cowardice on the one hand and the vice of foolhardiness on the other. We can think of the virtue of generosity as a middle ground between the vice of miserliness on the one hand and the vice of being a spendthrift on the other. We can think of the virtue of loyalty to an employer as a middle ground between the vice of complete disloyalty on the one hand and the vice of unquestioning obedience to the employer on the other. [2]

Various lists of virtues have been proposed, but there is a considerable overlap. The Greek philosopher Aristotle, the first and probably most important virtue ethicist, provided a very short list that includes courage, truthfulness, self-respect, wittiness, friendliness, modesty, and generosity or "magnificence." Contemporary psychologists Christopher Peterson and Martin Seligman have surveyed cultures throughout the world and come up with what they believe is a comprehensive list of core virtues and associated "character strengths", which are listed below.

1. Wisdom (creativity, open-mindedness, perspective)
2. Courage (bravery, persistence, vigor or energy)
3. Humanity (love, kindness)
4. Justice (citizenship, fairness, leadership)
5. Temperance (modesty, self-control)
6. Transcendence (appreciation of beauty and excellence, gratitude, spirituality) [3]

**Common morality**

In order to resolve some moral issues—especially those involving larger social policies—we must look more deeply into the moral ideas that lie at the basis of our moral judgements. The most obvious place to look is the stock of common moral beliefs that most people accept, called common morality.

The first account is by philosopher W. D. Ross, who constructed a list of basic duties or obligations, which he called "at first sight" duties. In using these terms, Ross intended to convey the idea that although any given duty is usually obligatory, it can be overridden by another duty in special circumstances. His list of at first sight duties can be summarized below:

1. Duties of fidelity (to keep promises and not to tell lies) and duties of reparation for wrong done.
2. Duties of gratitude (e.g., to parents and benefactors)
3. Duties of justice (e.g., to support happiness in proportion to merit)
4. Duties of beneficence (to improve the condition of others)
5. Duties of self-improvement
6. Duties not to injure others

Everyone recognizes that moral percepts and rules have exceptions. We have already seen that Ross calls his duties "at first sight", but he does not explain how we go about deciding when an exception is justified. There are two ways.

First, when moral duties or rules conflict, we must decide which percept is more binding in a given situation. Usually, it is wrong to lie, but if the only way to save an innocent person from being murdered is to lie to the assailant about the person's whereabouts, then most would agree that lying is justified. If we are willing for this exception to be widely practiced—if we are willing for others also to lie in similar circumstances—then the exception is justified.

Second, exceptions can be made to moral duties or rules when our conscience intervenes. Suppose a young man is called into military service, but sincerely believes that killing others is wrong, even to defend one's country. He might understand the obligation not to kill in war to follow from a duty not to injure others. The moral justification is that if this young man is forced to defend his country by killing, he is being forced to do what he deeply and sincerely believes to be wrong, and according to common morality as ordinary understood, one should never do what he or she deeply and sincerely believes to be wrong.

**Keywords**: Engineering ethic; Profession morality; Duties; Virtue

**Source from**:
Charles E. Harris, Jr., Michael S. Pritchard, Ray W. James. *Engineering Ethic: Concepts and Cases*. Sixth Edition. Boston: Cengage Learning, 2017.

# Part 3　Extension

## 1. *Engineering*(工程)

广义的工程概念认为：工程是一群人为达到某种目的，在一个较长的时间周期内进行协作活动的过程，强调众多主体参与的社会性。狭义的工程概念则认为，工程是以满足人类需求的目标为指向，应用各种相应的知识和技术手段，调动多种自然和社会资源，通过一群人的相互协作，将某些现有实体汇聚并建造为具有预期使用价值的人造产品的过程。狭义的工程概念不仅强调多主体参与的社会性，还跟生产实践密切联系，是运用一定的知识和技术得以实现的人类活动。

## 2. *Profession*(职业)

职业是参与社会分工，利用专门的知识和技能，为社会创造物质财富和精神财富，获取合理报酬，作为物质生活来源，并满足精神需求的工作。

职业是人类在劳动过程中的分工现象，它体现的是劳动力与劳动资料之间的结合关系，其实也体现出劳动者之间的关系，劳动产品的交换体现的是不同职业之间的劳动交换关系。

# Part 4　Words and Expressions

| | |
|---|---|
| pharmacy | *n.* 药房；配药学，药学；制药业；一批备用药品 |
| autopilot | *n.* 自动驾驶仪 |
| autonomous | *adj.* 自治的；有自主权的；[生，植]自发的 |

续表

| | |
|---|---|
| expertise | n. 专门知识或技能；专家的意见；专家评价，鉴定 |
| registration | n. 登记，注册；挂号 |
| penalty | n. 惩罚；刑罚；害处 |
| contemporary | adj. 当代的，现代的；同时代的，同属一个时期的<br>n. 同代人；同辈人；同龄人；当代人 |
| generosity | n. 慷慨，大方；宽容或慷慨的行为 |
| spendthrift | n. <贬>花钱无度的人，挥霍者；败家子<br>adj. 浪费的，奢侈的 |
| wittiness | n. 机智，临机应变 |
| magnificence | n. 华丽，富丽堂皇 |
| temperance | nj. 节制（尤指饮食），节欲；戒酒 |
| transcendence | n. 超越；超绝 |
| fidelity | n. 忠诚，忠实；逼真；保真度；尽责 |
| military service | n. 兵役 |

## Part 5  Reading

### Case

As engineers our work usually has a tangible effect on society by nature. We must therefore make certain that we follow a high code of ethics as the consequences of our actions can be immense. For example, the unethical constructor uses inferior materials during the construction process, resulting in insufficient load and breakage during the use of the bridge, which may cause casualties.

The typical event such as the Titanic, caused a large number of deaths due to the unethical decisions made by the Titanic's operators and the outdated safety regulations.[5] The Titanic was built to be the largest, safest, most opulent of its time and was supposed to be "unsinkable". However, it sank on its maiden voyage for hitting the iceberg. The lifeboats of the RMS Titanic played a crucial role in the disaster of 14-15 April, 1912. One of the ship's legacies was that she had too few lifeboats to evacuate all those on board. The 20 lifeboats that she did carry could only accommodate 1,178 people, despite the fact that there were approximately 2,208 on board. RMS Titanic had a maximum capacity of 2,220 passengers and crew.

Many lifeboats only carried half of their maximum capacity; there are many versions as to the reasoning behind half-filled lifeboats. Some sources claimed they were afraid of the lifeboat buckling under the weight, others suggested it was because the crew were following strict maritime tradition to evacuate women and children first. Few men were allowed into the lifeboats on the port side, while the starboard side only allowed men into boats after women and children boarded. Some of the final lifeboats were overfilled, and passengers noticed the seawater line was near the gunwale on some of the lifeboats.

Compounding the disaster, Titanic's crew were poorly trained on using the davits (lifeboat launching equipment). As a result, boat launches were slow, improperly executed, and poorly supervised. These factors contributed to the lifeboats departing with only half capacity. Besides, the Titanic did not have enough lifeboats to evacuate everyone on board. She only had enough lifeboats to accommodate approximately a third of the ship's total capacity. The shortage of lifeboats was not due to the lack of space and nor was it because of cost. At last, 1,503 people did not make it on to a lifeboat and were aboard Titanic when she sank to the bottom of the North Atlantic Ocean at 2:20 a.m. on 15 April 1912. If the designers have made more ethical decisions and the crews were skilled in mastering professional skills, the casualties would not be so heavy.

**Reference:**
[1] Davis M. Is there a profession of engineering? [J]. Science and Engineering Ethics, 1997, 3(4):407-428.
[2] W. D. Ross, Aristotle. London: Methuen, 1930: 203.
[3] Christopher Peterson and Martin E. P. Seligman, character strengths and virtues. New York: Oxford University Press, 2004: 29-30.
[4] Li Zhengfeng, Cong Hangqing. Engineering Ethic[M]. Beijing: Tsinghua university press, 2016.
[5] Berg, Chris. "The Real Reason for the Tragedy of the Titanic". Wall Street Journal, 2012-4-13.

**Reference:**
维基百科。

## Part 6　Exercise and Practice

### 1. Discussion

When moral duties or rules conflict, how to decide which percept is more binding?

### 2. Translation

(1) The issues are important to engineers not simply because engineers have usually created the technologies involved, but because engineers are professionals, and the concept of professionalism has a strong moral component.

(2) A profession is a number of individuals in the same occupation voluntarily organized to earn a living by openly serving a moral ideal in a morally permissible way beyond what law, market, morality, and public opinion would otherwise require.

(3) Someone may claim that professional codes are mere window dressing, designed to disguise the fact that professionals are primarily out to promote their own economic self-interest.

(4) We can think of the virtue of loyalty to an employer as a middle ground between the vice of complete disloyalty on the one hand and the vice of unquestioning obedience to the employer on the

other.

(5) Usually, it is wrong to lie, but if the only way to save an innocent person from being murdered is to lie to the assailant about the person's whereabouts, then most would agree that lying is justified.

## 3. *Writing*

Explain the measures you will take if engineering ethics conflict with personal interests and even personal safety.

# 附 录

## 附录1　国际专业学术团体

### IIR

(International Institute of Refrigeration)

国际制冷学会

　　国际制冷学会(IIR)是一个独立的政府间交流科学和技术的国际学术组织。它致力于推动企业、实验室以及组织内的发展进程,传播制冷及相关技术方面的先进知识,以提高人类的生活质量;提升科学和技术领域内所涉及的制冷技术及其应用,为人类谋求福利;促进和扩展国际合作,为业内人士在全球范围提供市场与技术导向的高质量服务。

　　目前,国际制冷学会有61个会员国,这些会员国的人口约占全世界的80%。我国于1978年1月加入该会,为二级会员国。此外,企业、实验室、大学等也可成为国际制冷学会的企业及捐助人成员。

　　国际制冷学会的出版物有:《国际制冷学会通报》(双月刊)、《国际制冷杂志》(双月刊),以及各种推荐条件和规程、各类小册子、大会及专业委员会学术会议论文集等。

### ASHRAE

(American Society of Heating, Refrigerating and Air-conditioning Engineers)

美国采暖、制冷与空调工程师学会

　　美国供热、制冷与空调工程师学会成立于1894年,是一个拥有50 000多个会员,分会遍及全球的国际性组织。该学会聚焦建筑系统、能源效率、室内空气品质和可持续发展等行业领域。通过开展科学研究,提供标准、准则、继续教育和出版物,促进供热、通风、空调和制冷方面的科学技术发展。

该学会作为行业的全球领导者,旨在全面推进供暖、通风、空调及制冷的科学技术,造福社会公众,促进世界的可持续发展。

## CIBSE

(The Chartered Institution of Building Services Engineers)
英国皇家屋宇装备工程师学会

英国皇家屋宇装备工程师学会于1976年获得皇家特许。它致力于"为会员和公众提供一流的信息和教育服务,并推动团队精神指导工作,以服务工程建设的科学技术发展和实践"。

该学会促进设备工程师的认证与学习,提供一系列的服务,协助获得资质的设计师在整个职业生涯中保持和提高专业上的卓越。该学会还是工程建设标准的制定者和权威,其出版的指引和守则,是国际公认的权威,在行业中作为最佳实践的标准。

## SHASE

(The Society of Heating, Air-conditioning and Sanitary Engineers of Janpan)
日本空气调节和卫生工程学会

日本空气调节和卫生工程学会是日本在采暖、空调和卫生工程领域重要的学术组织,已超过90年的历史。

该组织成立于1917年,迄今已有超过20 000名会员。学会通过教育、出版、科学研究、国际合作、编制标准、组织资格考试、设立奖项等形式,大力促进日本建筑设备与环境工程的科学技术发展和应用。

## IGEM

(Institution of Gas Engineers and Managers)
英国燃气专业学会

英国燃气专业学会是经英国特许工程委员会授权的专业机构,通过发展会员、组织活动和编制技术标准,广泛服务英国和国际的天然气行业。

学会的会员关系能满足天然气行业的所有成员需求,从大学生到合格的专业人员,再到有兴趣参与其中的社区人士,都可以通过学会找到合适的位置。英国燃气专业学会每年会举办各式各样的活动,并由此产生了大量的技术标准。

## 附录 2　国际专业学术期刊

**Building and Environment**：建筑科学与应用领域的国际期刊，主要发表与建筑科学、建筑环境以及在建筑设计和应用中的人机交互、技术和工具开发等有关的研究论文。

**Energy and Buildings**：能源应用及建筑能效领域的国际期刊，主要发表与建筑能源利用相关的文章。其目标是展现新的研究成果，展示减少能源需求和改善室内环境质量的、新的、行之有效的做法。

**Indoor Air**：室内环境与空气品质领域的国际期刊，主要发表在非工业建筑中与室内环境相关的原创性研究成果，旨在让设计师向业主和运营商提供一个健康、舒适的环境，并向医务人员提供关于如何处理与室内环境相关的疾病的信息。

**ASHRAE Journal**：ASHRAE 的会刊之一，主要发表当前暖通空调制冷行业面向应用的文章。期刊还发表关于新兴技术的评论，文章内容覆盖从设计到施工再到运行调试的整个系统生命周期。

**HVAC&R Research**：ASHRAE 的会刊之一，主要提供关于建筑环境调控和冷却应用技术领域的研究报告，并涉及热力学、流体力学和传热学等专业基础领域的研究论文。

# 附录3　英文摘要写作

摘要作为论文的重要组成部分，是读者快速获得全文主要信息的关键途径。摘要比论文全文具有更广泛的读者。一方面读者会根据摘要提供的信息，做出是否阅读全文的判断。另一方面，英文摘要作为信息检索的工具，直接影响论文的被检索次数和被引率。此外，编辑人员也会通过摘要大致评估论文的深度和创新性，从而缩短选稿进程。

英文摘要是全文的浓缩，概括了论文从引言（introduction）、方法（methodology），到结果（results）、结论（conclusion）每一部分内容的要点，可独立存在。英文摘要一般不超过250个英文单词，应尽可能简明、完整、规范。一般情况下，摘要中不列图、表，不引用文献，不加评论和补充解释。在英文摘要中，经常使用的时态是一般现在时和一般过去时。具体用何种时态，应根据表达的内容而定。一般现在时多用于说明研究目的、描述研究内容、结果、结论等；一般过去时多用于叙述过去某一时刻（时段）的研究过程及其发现等。因为科技论文更强调说明对事物的研究过程，因此英文摘要一般多用被动语态。但为了有助于文字清晰和简洁，现在主动语态的采用也越来越多。摘要中出现的缩略语、代号等，除了公知公认者外，首次出现时须注明全称。

**What are the purpose of abstracts**?

What are the uses of abstracts?

——Abstracts should give complete and detailed short summaries of technical articles.

（1）Professionals use abstracts to survey publications and recent research in their field of speciality.

（2）Professionals use abstracts to determine which publications they will read.

（3）Editors and conference organizers use abstracts to determine whether or not a paper will be accepted for publication or accepted for presentation at a conference.

**How long should an abstract be**?

One or two paragraphs. Generally abstracts have a maximum of 250 English words.

**What should an abstract contain**?

（1）A one sentence statement of the main point of the paper.

（2）A description of the investigation, experiment, procedure, or equipment of interest.

（3）Indication of the results, findings, or data obtained.

（4）The conclusions that can be drawn from the experimental results, findings or data.

（5）A statement of the implications of the conclusions drawn for further experiments, applications, or new topics of research. Negative results are often published to prevent needless duplications of research efforts.

### How are abstracts judged?

On the basis of:

(1) Technical significance

(2) Specificity

(3) Precision

(4) Completeness and comprehensiveness

(5) Economy of words

—What happens if an abstract is overgeneralized, wordy, imprecise, or incomplete?

(1) Professionals will not bother to read the entire paper.

(2) Editor and conference organizers will reject the paper.

### What style is right for abstracts?

—Because abstracts are so short, any errors will be very noticeable. The following rules are important in all writing. They are especially important when writing abstracts:

(1) State the main point of the paper in a single sentence.

(2) Organize your thoughts. State main ideas first, then give examples.

(3) Shorter is better. Don't waste words. Make every word count.

(4) Judge your audience.

    (a) Don't tell your readers what they already know.

    (b) Don't assume your readers know things they do not.

(5) Avoid cliches and stock phrases.

### Examples

1. The coupling between heat pumps and renewable energy sources is a recognized strategy to reduce primary energy consumption at the household level. This paper contributes to the present-day discussion concerning solar-assisted heat pumps for heating/cooling and to produce domestic how water. In particular, this study presents the results of a field study concerning a novel solar-assisted dual-source multifunctional heat pump, installed in a detached house in Milan. The proposed system couples hybrid photovoltaic/thermal (PVT) panels with a multifunctional and reversible heat pump. The heat pump is equipped with "air-source" and "water-source" evaporators, connected in series and operated alternatively, based on the ambient conditions, system parameters and operating modes. In addition, the PVT panels are used, by employing two storage tanks, to produce domestic hot water and to provide a heat source to the "water-source" evaporator. The proposed system has been tested experimentally, showing interesting and promising results: the system has been able to maintain high efficiencies in the different seasons and has been able to use the solar energy to support the production of domestic hot water. It was found that the use of the "water-source" evaporator allowed to significantly increase the performance of the system and to avoid the defrost cycles.

2. Due to their non-deterministic behaviour, renewable energies are defined as non-dispatch-

able energies and are largely coupled with energy storage systems to overcome the problem of matching energy production and demand. Hence, in the energy efficiency and conservation field there is growing interest towards energy storage systems, especially when combined with the demand side management (DSM) concept, representing DSM the possibility of shaping end user electricity consumption. In this work an existing installation of a thermal energy storage (TES) system coupled with heat pumps in an industrial building is presented and a dynamic simulation model is built to represent its behaviour. Simulations are performed to show the load shifting potential of such storage and costs and energy use are assessed for different configurations, in order to evaluate the viability of this TES application. In particular the demand side strategy considered is aimed at shifting energy demand for cooling to weekend daytime to recover surplus PV electricity or otherwise to off peak hours to profit from lower electricity tariffs. It is found that the use of TES implies increased energy demand, while costs can decrease when electricity tariffs with a considerable difference between on peak and off peak rates are applied. Furthermore the integration with renewable sources for electricity production, such as PV panels, makes the installation of TES economically interesting independently of the electricity tariff in place. However the more relevant aspect for the overall economic feasibility of such installation is the initial capital investment.

## 附录4 科技论文翻译(中文)

### 1. 翻译原则

中国早期的翻译家严复提出翻译的活动标准为"信、达、雅"的思想。"信"(faithfulness):忠于原文;"达"(intelligibility):表达原文主旨,译文通顺,文以载道;"雅"(elegance):译文雅典得当。科技译文追求"信、达、雅",就要求首先对原文的科技内容有本质的了解,在意译的基础上通过中英文的转换传递信息,并重现原文风格。

而西方的翻译理论中有美国学者提出核心思想为"功能对等",即为实现译文与原文在语言功能上的对等,而非形式。为使译文自然,必须摆脱原文语言结构的束缚,由于英汉两种语言结构不同,在翻译时有所改变是必然的。Eugene A. Nida 指出为抛开原文语法、语义结构的影响,可采用如下方式:a. 用词类取代传统词性来描述语义关系;b. 用核心句或者句型转换概念克服句法障碍;c. 使用同构体的概念科学地实现翻译语言功能的对接。

### 2. 译文的特点

科技语言有自身的特点,不同于社会科学受到本国文化传统影响较大,现代科学更倾向于全球一体化,在翻译时应该相应地使用译入语的专有词汇和表达结构,以便于读者迅速理解文章的意思。科技论文的译文应当重视文章内容,而不是形式,实现功能对等,应当尽量保持译入语的特色。

### 3. 中译英

#### 3.1 标题翻译

论文的题目是对全文的高度概括和总结,既要体现文章的内容,又要吸引读者,因此标题文字要求精练、准确和醒目。中文论文题目必须简洁明了、准确,而将中文题目翻译成英文时,既要把原标题的意思完整表达,又要使之符合英语用词习惯。翻译时,应先了解文章的内容,正确理解专业名词的翻译,英文题目一般以短语为主,先明确中心词,再加修饰词。

##### 3.1.1 标题翻译举例

不定式短语:"怎样提高学生的实践能力"翻译成:How to Improve the Students' Practical Ability。

动名词短语:"强化过程控制 降低费用支出"翻译成:Reducing enterprise cost and charge by strengthening management process control。

名词短语:"齿刀是怎样演变成齿轮滚刀的"翻译成:Transformation from Rack-type Geer Cutter into Geer Hobber。

介词短语:"浅谈多媒体技术"翻译成:About Multimedia Techniques。

##### 3.1.2 标题翻译注意事项

翻译中文题目需要注意的是,防止逐词死译,恰当使用构词,注意冠词的取舍等。翻译前熟读文章,仔细推敲标题的含义,理解逻辑关系,明确专业术语的正确表达,不要生搬硬套。

应该恰当地使用构词,标题翻译常用构词成分来表达某些意思,这样能使标题更为简洁。如"multi-"(多元的),"fore-"(预先),"trans-"(跨越),"poly-"(多、复),"sub-"(分、再),等等。

例如:"多中段开采分段崩落法通风系统的建立"翻译为:Establishment of the Ventilation System of Sub-level Caving in Poly-level Mining(用了 sub-和 poly-)。

注意冠词的取舍,在论文标题的翻译中,有时省去冠词能突出重点,但在有些情况下,保留冠词,能使意思更加明确。例如:"基于数据包络分析的空调系统控制策略评估"翻译为:Control strategy evaluation of air conditioning system based on data envelopment analysis(省略冠词);"墙壁开洞爆破技术"翻译成:Techniques for Blasting through a Hole in the Wall(保留冠词)。

## 3.2 摘要翻译

摘要又称为概要、内容提要,目的是提供文献内容的整体梗概,不加任何评论和补充解释,简明、确切地记述文献重要内容的短文。科技论文摘要构成的五个要素包括背景、研究目的、方法、结果和结论。

背景:介绍研究的背景、现状和问题。

目的:说明研究的前提、目的、任务以及所涉及的主题范围。

方法:介绍研究所使用的原理、对象、材料、工艺、程序等。

结果:说明研究的结果、效果、性能等。结论表明对研究结果的分析、比较和应用,或者根据结果提出的问题、建议以及预测等。

下面分别举例《公共建筑节能改造技术途径与效果分析》《侧墙和顶板毛细管辐射供冷性能的实验研究》两篇论文的摘要翻译。

《公共建筑节能改造技术途径与效果分析》中文摘要:

基于123个既有公共建筑节能改造工程,总结梳理了既有公共建筑节能改造技术体系。统计了各类改造技术所取得的节能效果,分析了节能改造工程中影响节能效果、改造质量的关键因素,并提出了相关工程建议,所提出的技术体系,以照明插座系统和空调系统改造为核心,动力系统、供配电系统、特殊用电的系统改造为辅助,可实现约21%的总节能率。基于实际工程经验,重点讨论了照明改造中光源质量与二次照明设计存在的问题,探讨了目前集中空调水系统变流量改造的实施现状与限制,并总结了各项节能改造技术的适用性和节能效果。

英文摘要:

Based on 123 energy efficiency retrofitting projects, summarizes the energy efficiency retrofitting technical system of existing public buildings and the energy efficiency effect of in energy efficiency retrofitting technology, analyses the factors influencing energy efficiency effect and retrofitting quality, and proposes corresponding engineering suggestions. The proposed technical system with the lighting and air conditioning system retrofitting as core, dynamical system, power supply system and special energy using system retrofitting as assistance, can achieve 21% of overall energy efficiency rate. Based on the real projects, discusses lamp quality and secondary design of illumination, current situation and limitation of variable-flow retrofitting on central air conditioning system, and summarizes applicability and effect of each energy efficiency technology.

《侧墙和顶板毛细管辐射供冷性能的实验研究》中文摘要:

研究了供水温度16 ℃和18 ℃时侧墙和顶板毛细管辐射供冷室内热环境响应特性及舒适性等。结果表明:两种末端供冷室内平均辐射温度和操作温度均在4 h内达到稳定,且稳定阶段温度低于28 ℃;毛细管间距2 cm和4 cm情况下侧墙辐射单位面积热流量稳定值差20.62 W/m²。稳定阶段两种辐射末端室内温度分布均匀,顶板辐射室内满足Ⅰ级热环境标准

（PMV≤0.5），侧墙辐射满足Ⅱ级热环境标准（0.5＜PMV≤1）。两种形式辐射末端都能营造舒适室内热环境。

英文摘要：

Analyses the indoor thermal environment response and comfort of wall and ceiling capillary radiant cooling at the system water supply temperature of 16℃ and 18 ℃, respectively. The results show that the indoor thermal environment of both terminals can be stabilized within 4 hours and both the temperature in steady-state are kept at 28 ℃. The difference of heat flow per unit area of wall radiant cooling is up to 20.62 W/m$^2$ for the capillary tube distance of 2 cm and 4 cm. The indoor vertical and horizontal temperature distribution of both terminals are uniform, and the thermal environments of wall and ceiling radiant cooling systems are evaluated as Grade Ⅱ (0.5 < PMV ≤ 1) and Grade Ⅰ (PMV≤0.5) in steady-state, respectively. The indoor thermal environment of two radiant terminals can meet the comfort requirements of the standard.

### 3.3 翻译注意事项

翻译时，需要按照翻译的语言对应的思维去表述。英文科技论文翻译中常见的错误有语法、语义错误，句子结构和介词搭配错误、时态错误以及其他一些错误。

语法和语义的错误主要表现在单复数、人称、时态以及词语的表达含义理解问题。单复数是英文表达的特色，有些词既是可数名词又是不可数名词，这时应当注意该词在文中的含义，例如：influence、think、study、research 等。同时谓语动词应当与主语的单复数保持一致。对于语态的使用，全文并非保持被动语态不变，应当根据具体情况进行适当变化。例如"It can be seen from the results"可修改为"The results indicate"。在翻译时，应当考虑上下文的一致，而不必拘泥于词语上的一致。比如"科学技术"是由"科学"和"技术"二词并列组成，不是修饰关系，翻译为"scientific technology"不恰当，应翻译为"scientific and technology"。

举例说明句子结构不当引起的语义错误：It proves that by using new type drainage controller, the effect is better, it has higher safety and economy。（实践证明，采用新型排水调节器，效果更佳，具有较高的安全性及经济性。）修改为：The new type drainage controller has proved more effective, safer and more economical。

应当注意的是，英文中的标点符号与中文有差异：

①英文中没有"《》"和"、"，英文中书名号使用斜体书写，中文中的顿号在英文中可用逗号代替。

②英文中的省略号为"..."，且位置靠下，中文省略号为"……"。

③汉语的标点符号前后均不用空格，但在英文中，除破折号前后无须空格外，大多数标点符号后面都有空格。

## 4. 英译中

### 4.1 翻译技巧

有学者提出英语科技文献翻译是一个文献再创作的过程，在遵循原文意思的基础上，理清文献的整体脉络，翻译出意思明确的文献资料。英文科技论文的翻译应当做到：一是把原文变成另外一种文字时不能改变原作者所表达的意思；二是做到不增、不减、不改。例如：It is reported that 2 million people have been inoculated. Then blood slides are examined with microscope.

翻译为:据报告已有二百万人接受注射。血片已用显微镜加以检查。而不是翻译为:据有关报道……(这样就在原文基础上增加了多余的信息)。

对于英译中,科技论文与文学作品不同,科技论文用词俭朴,易于理解,很复杂的句子和字比较少见,但是科技论文中的普通字,常有特殊的解释。对普通单词应进行专业术语翻译。例如"Substation"翻译为"热力站";"Primary/Secondary(side)"翻译为"一次网(侧)/二次网(侧)";"Feed(Supply,Flow)/Return"翻译为"供水/回水";"Water softener"翻译为"水处理装置"。

另外,对于翻译者必须掌握英文的语法知识,明确句子的语法结构,才能正确地翻译。必要的时候借助工具书或相关翻译软件进行翻译。

## 4.2 长句翻译

英语长句在整体结构上具有自身的特点,主要表现为在句型结构较为复杂的语法和句型相互交叉,使得整个句式翻译难度增加。文献中出现的长句主要包括主从复合句、并列复合句、简单长句。而对于长句的翻译技巧主要有结构分析法、拆分法、顺译法、变序法、重组法。结构分析法即对句子结构进行分析,其次明确句中连词的使用,最后对翻译后的句子表达习惯进行确定,保证其符合语言习惯的要求。拆分法即将英语中复杂的句子拆分成几个较为简单的句式,进行翻译而后重组。顺译法主要是指结合文章中的逻辑关系,对句子进行直接的翻译。当顺着原文的顺序翻译成汉语时不够通顺,以致难以理解时,可采用变序法,需打乱原文顺序,做不同安排。句子结构的重组一般发生在英文语序与中文语序之间不一致时,中文的逻辑与英文存在差异,在翻译过程中将句子的整体语序进行调换。

# References and Sources

[1] 张寅平,潘毅群,王馨. 高等学校专业英语系列教材——建筑环境与设备工程专业[M]. 北京:中国建筑工业出版社,2005.

[2] 赵三元,闫岫峰. 建筑类专业英语——暖通与燃气. 第一册[M]. 北京:中国建筑工业出版社,2002.

[3] 何天祺. 供暖通风与空气调节[M]. 3版. 重庆:重庆大学出版社,2014.

[4] 孙一坚,沈恒根. 工业通风[M]. 4版. 北京:中国建筑工业出版社,2010.

[5] 姚杨. 暖通空调热泵技术[M]. 北京:中国建筑工业出版社,2008.

[6] 付祥钊. 建筑节能原理与技术[M]. 重庆:重庆大学出版社,2008.

[7] 林崇德. 心理学大辞典[M]. 上海:上海教育出版社,2003.

[8] ASHRAE. *ASHRAE Handbook* 2007, *HVAC Applications*. Atlanta, GA:ASHRAE, 2007.

[9] ASHRAE. *ASHRAE Handbook* 2008, *HVAC Systems and Equipment*. Atlanta, GA:ASHRAE, 2008.

[10] ASHRAE. *ASHRAE Handbook* 2009, *Fundamentals*. Atlanta, GA:ASHRAE, 2009.

[11] ASHRAE. *ASHRAE Green Guide*:*The Design*, *Construction*, *and Operation of Sustainable Building*, Atlanta, GA:ASHRAE, 2010.

[12] CIBSE. *Guide B Heating*, *ventilation*, *air conditioning and refrigeration*. London:CIBSE Publications, 2005.

[13] IGEM. *Gas Legislation Guidance IGE/GL/9 Communication 1724*. (Guidance for large gas consumers in dealing with Natural Gas supply emergencies), 2006.

[14] Roger W Haines, Michael E Myers. *HVAC System Design Handbook*. New York:McGraw-Hill, 2010.

[15] Shan K. Wang. *Handbook of Air Conditioning and Refrigeration*. 2nd ed. New York:McGraw-Hill Professional, 2000.

[16] W. Whyte. *Cleanroom Technology-Fundamentals of Design*, *Testing and Operation*. England:John Wiley & Sons Ltd, 2001.

[17] Dong Xiucheng. Origins and countermeasures for "gas famine" in China. *Natural Gas Industry*, 2010,30:119-122.

[18] Liang Junyi. Key Points of Natural Gas Purification Process Design and Their Optimization. *Natural Gas and Oil*, 2012,30:32-35.

[19] Wu Xuehong. Operation Management of Underground Natura Gas Storage with Salt Caverns. *Natural Gas and Oil*, 2008, 26:1-4.

[20] Wang Tao. Application of Entropy Technology in Safety Assessment on Natural Gas Gathering and Transportation Station. *Natural Gas and Oil*, 2012, 30:10-13.

[21] Yang Fan. Discussion on Long-distance Gas Pipeline Design in New Situation. *Natural Gas and Oil*, 2012, 30:14-16.

[22] Ruud Weijermars. *Journal of Natural Gas Science and Engineering*, 2010, 2: 86-104.

[23] 李原. 浅议中文科技论文的摘要和英文文摘[J]. 编辑学报,1998,10(2):97-99.

[24] 叶子南. 高级英汉翻译理论与实践[M]. 北京:清华大学出版社,2001.

[25] 邓清燕. 科技论文英文文摘的翻译原则与常见错误分析[J]. 北京工商大学学报:自然科学版,2004(04):64-68.

[26] Dark. 什么是摘要[EB/OL]. 2012-10-28[2019-03-06]. http://iask.Sina.com.cn/b/1964328.html.

[27] 冯志杰. 汉英科技翻译指要[M]. 北京:中国对外翻译出版公司,2000.

[28] 唐浩,丁勇,刘学. 公共建筑节能改造技术途径与效果分析[J]. 暖通空调,2018,48(12):118-125.

[29] 杜晨秋,李百战,刘红,等. 侧墙和顶板毛细管辐射供冷性能的实验研究[J]. 暖通空调,2018,48(05):103-110.

[30] 赵春娥. 谈科技论文题目、摘要及关键词的英文翻译[A]. 山西省科技情报学会2004年学术年会论文集[C]. 山西省科技情报学会,2005:5.

[31] 李晴. 翻译技巧在英语科技文献中的运用研究[J]. 海外英语,2014(01):118-119.

[32] 范琴英. 科技论文翻译浅谈[J]. 中国科技信息,2005(15):270.

[33] 罗月丰. 当代大学生思想政治状况调研[J]. 思想政治教育,2004,13(1):40-45.

[34] 李娜. 英语科技文献中长句翻译技巧[J]. 智库时代,2017(10):278-279.